Introduction to
ECOTOXICOLOGY

Des Connell School of Public Health,
Griffith University, Queensland, Australia

**Paul Lam, Bruce Richardson
and Rudolf Wu**
Department of Biology and Chemistry, City University of Hong Kong

Blackwell
Publishing

© 1999 by
Blackwell Science Ltd
Editorial Offices:
Osney Mead, Oxford OX2 0EL
25 John Street, London WC1N 2BL
23 Ainslie Place, Edinburgh EH3 6AJ
350 Main Street, Malden
  MA 02148 5018, USA
54 University Street, Carlton
  Victoria 3053, Australia
10, rue Casimir Delavigne
  75006 Paris, France

Other Editorial Offices:
Blackwell Wissenschafts-Verlag
  GmbH
Kurfürstendamm 57
10707 Berlin, Germany

Blackwell Science KK
MG Kodenmacho Building
7–10 Kodenmacho Nihombashi
Chuo-ku, Tokyo 104, Japan

First published 1999

Reprinted 2004, 2005

Set by Graphicraft Limited,
  Hong Kong

Printed and bound in
Great Britain by
TJI Digital, Padstow, Cornwall

The Blackwell Science logo is a
trade mark of Blackwell Science Ltd,
registered at the United Kingdom
Trade Marks Registry

A catalogue record for this title is
available from the British Library

ISBN 0-632-03852-7

Library of Congress
Cataloging-in-publication Data
Introduction to ecotoxicology / by
Des Connell . . . [et al.].
    p. cm.
    ISBN 0-632-03852-7
    1. Pollution–Environmental
aspects. I. Connell, D. W.
    QH545.A1I5745   1999
    577.27—dc21   98-53192
                        CIP

DISTRIBUTORS

Marston Book Services Ltd
PO Box 269
Abingdon, Oxon OX14 4YN
(*Orders*: Tel: 01235 465500
        Fax: 01235 465555)

USA
Blackwell Science, Inc.
Commerce Place
350 Main Street
Malden, MA 02148 5018
(*Orders*: Tel: 800 759 6102
        781 388 8250
        Fax: 781 388 8255)

Canada
Login Brothers Book Company
324 Saulteaux Crescent
Winnipeg, Manitoba R3J 3T2
(*Orders*: Tel: 204 837-2987)

Australia
Blackwell Science Pty Ltd
54 University Street
Carlton, Victoria 3053
(*Orders*: Tel: 3 9347 0300
        Fax: 3 9347 5001)

For further information on
Blackwell Science, visit our website:
www.blackwell-science.com

Contents

Preface

Ecotoxicology is becoming more prominent in the curricula of wildlife, biology and environmental science departments in universities and colleges. This book is a response to that development and addresses some of the aspects which have become more significant over recent years such as ecological risk assessment and biomarkers. However, it takes a broad view of the subject ranging from the chemical and biochemical level to ecosystems and management.

The book is aimed principally at undergraduate students who have completed basic courses in biology and chemistry. These students could be in areas such as biology, ecology, wildlife management, environmental sciences, toxicology, chemical engineering, civil engineering, sanitation engineering and other similar disciplines. The book will also prove useful to consultants, civil servants and other people involved in water quality evaluation, waste engineering, environmental impact assessment, biological conservation, toxic chemical management and so on.

Ecotoxicology has had a relatively short gestation period. The gathering of information regarding chemicals in the environment first commenced during the 1950s with the development of sensitive techniques for the analysis of chemical residues in natural systems. A picture gradually emerged during the 1960s of the widespread but low concentrations of pesticides in birds, fish and other organisms. The book *Silent Spring* by Rachel Carson (1962) attributed a variety of deleterious effects to these persistent agricultural chemicals. This stimulated a major research effort into the effects of chemicals in natural ecosystems which continues to the present day.

The evaluation of these chemicals in natural systems has posed a set of problems not previously encountered in investigations of toxic chemicals. Prior to this period toxicology was principally concerned with lethal or therapeutic doses of substances for mammals or pest organisms with special relevance to human situations. On the other hand the chemicals in the environment belong to relatively persistent chemical groups, usually occur in extremely low concentrations and in many situations do not have immediate lethal or therapeutic effects; and the organisms of concern belong to many different biotic groups apart from mammals and pests. It has become apparent that a new approach is needed to understand chemicals in the environment. This has led to the development of 'ecotoxicology' which essentially provides a broad conceptual framework for evaluating chemicals in the environment.

The use of the term ecotoxicology has not yet found universal agreement. Some see it as the evaluation of the toxicity of chemicals to organisms in natural systems. In this book ecotoxicology has been taken to mean the study of the pathways of exposure, uptake and effects of chemical agents on organisms, populations, communities and ecosystems.

Many people assisted us in the preparation of this book. The original concept was suggested by Simon Rallison and developed by Ian Sherman of Blackwell Science. The preparation of the final complex manuscript was capably carried out by Rahesh Garib. To all of these people we are extremely grateful.

Des Connell
Paul Lam
Bruce Richardson
Rudolf Wu

1: The Ecotoxicology Concept:
An Introduction

The essence of toxicology: a multidisciplinary approach

The term 'ecotoxicology' was first coined in 1969 by Professor R. Truhaut, who defined it as a science describing the toxic effects of various agents on living organisms, especially on populations and communities within ecosystems. The essence of ecotoxicology lies in two main areas: a study of the environment, with origins in the science of *ecology*; and a study of the interactions of toxic chemicals with individual living organisms—the science of *toxicology*. Humans have always interacted with their environment, relying upon it for a source of building materials, clothing, sustenance, etc. Toxicology has also played an important part in human history. For example, poisoned arrows have long been used to kill animals for food (a tradition continuing today in certain remote African and South American tribes). There are numerous other examples of toxicology in practice: from water pipes contaminated with lead in ancient Rome to the more modern horrors of the gas chambers during the Holocaust of World War II, the release of radioactive material at Chernobyl and the ever-present threats of chemical warfare in many parts of the globe.

Truhaut's definition of ecotoxicology has been followed by several others. Importantly, recent definitions have included two aspects that dominate modern approaches to the subject. The first of these includes the notion that hazards to living organisms from toxic chemicals can be investigated through the use of survey data; in other words, a *retrospective* approach, which assesses the levels of toxic chemicals in the environment and uses this information to determine their potential past, present and future impacts. In essence, this takes the form of a 'historical environmental perspective'—looking at a situation which *has* occurred, and linking the cause (i.e. the chemicals) with effect (i.e. the response of the living organism to these chemicals). The second aspect, in contrast, seeks to *predict* the impact of chemicals through *prospective* studies. This involves the use of specific tests which allow scientists to assess the *likely impact* of single chemicals or mixtures such as complex effluents and industrial and municipal wastes.

Ecotoxicology, by its very nature, must be multidisciplinary, combining the sciences of chemistry, toxicology, pharmacology, epidemiology and ecology with an understanding of the sources and fates of chemicals in the environment. To this must be welded a managerial aspect, resulting from the increasing need to regulate industrial and human activities which may possibly cause pollution. Thus, we can also enter the concepts of risk assessment and risk management into the ecotoxicological equation. These facets seek to

1

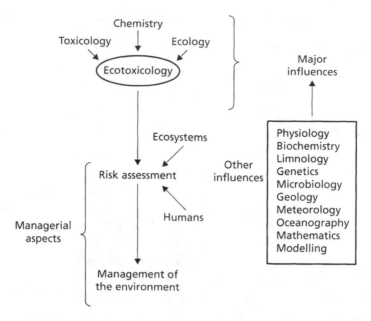

Fig. 1.1 The various components that contribute to *ecotoxicology*.

determine and assess the inputs and fates of chemicals (especially new chemicals) in the environment, and to manage the possible consequences of their introduction (Fig. 1.1).

Perhaps the best definition of ecotoxicology is that of Moriarty (1988), who suggests that the task of this science is to assess, monitor and predict the fate of foreign substances in the environment. This definition encompasses the chemical characterization of contaminants that are present in our environment; the need to monitor them in order to assess whether environmental loads are increasing or decreasing; and the need to predict their impacts through studies involving modern methods of toxicology and ecology.

A short history of the chemical age

At the present time, the number of chemicals routinely in use by human society is enormous. It is estimated that around 70 000 chemicals are commonly utilized for a wide variety of purposes in our global community, and that the rate of introduction of new substances is in the order of 200–1000 compounds per year. Indeed the manufacture of novel substances which do not occur naturally (the so-called *xenobiotics*) is a trend which has increased markedly during the present century.

All human communities throughout history have produced wastes. Such wastes included sewage, which was typically disposed of into water bodies (e.g. rivers, lakes and seas), or alternatively onto the land or in burial sites where natural degradation over time could occur. Historically, disposal in this manner was easy and convenient, as most communities were small and centred upon a location where a ready supply of water could be accessed—for instance, close to rivers and lakes.

Such disposal may be easy when communities are relatively small in size, but as human populations grew, and larger cities were built, waste disposal became an increasingly difficult problem, leading some communities to develop rules and regulations for proper handling of the material. For instance, laws in Athens around 500 BC prevented the disposal of rubbish within the city walls. The ancient Romans built aqueducts and sewers, and took steps to prevent the disposal of rubbish in their most important waterway, the River Tiber. These Greek and Roman laws may therefore be amongst the earliest that we would now recognize as legally enforceable environmental protection regulations.

It should be noted that the vast bulk of these wastes were natural substances based upon products of the environment. We recognize today that disposal of these wastes resulted in physicochemical and microbial-induced degradation, and that the wastes generated by relatively small community sizes could be readily handled by the receiving environments.

An enormous change resulted from the *Industrial Revolution*—a turning point for modern society in terms of ecotoxicological studies. In the late 18th century, human culture in many countries changed radically, shifting from a rural base towards an increasingly industrialized civilization. A demographic shift in population also occurred: large numbers of rural people shifted to the cities and industrial centres, where jobs, money and a more comfortable lifestyle were available. A greater supply of fuels, the result of more intensive mining activities, fired the factories: coal, generating steam power, combined with newly developed machines to speed production. The first recognizable ecotoxicological problems at this time were the result of intensive industrial activity in confined areas. Rural areas were rapidly converted to industrial centres, as factories were placed near to sources of fuel. Coal usage—the principal form of energy driving the industrial revolution—produced localized and regional problems of atmospheric contamination. A London doctor, Percival Pott, noted the high incidence of occupationally-induced cancer of the scrotum among chimney sweeps, many of whom crawled up through chimneys to do their job. He believed that this was the result of exposure to soot or, more precisely, a group of chemical compounds produced as a result of the combustion of fossil fuels—the polycyclic aromatic hydrocarbons, or PAHs.

In the early 1900s, another change was to take place which would eventually limit the use of coal to fewer industries. A new fuel—petroleum—was found and a whole new industry created, not only for its extraction and refining, but also for finding new ways to utilize this material and the new substances which could be formed from it. The 20th century rapidly became the age of oil. Indeed, most industries are dependent upon a form of petroleum-based substance for some or all of their activities.

The introduction of the motor car and petroleum-based fuels had other repercussions. As engines were refined, so too petrol was further improved. Antiknock fuels, containing lead, were first introduced in the 1920s, and produced widespread contamination of urban areas and localities near major traffic arteries throughout the rest of the century.

Another highly significant change was quietly taking place—a revolution in agriculture—to cope with the increasing population size. The development of fertilizers was perhaps the critical step which drove agricultural production forward. The important breakthrough—the manufacture of nitrogenous fertilizers from atmospheric nitrogen was developed by Fritz Haber in Germany with factories operating by 1913. With the widespread use of fertilizers began a 'chemical' age in agriculture, whereby poorer soils could be more effectively utilized for agriculture, large crop yields could be gained year after year, and plentiful supplies would allow exports to countries unable to produce sufficient in their own right. The net result was an increased human survival and a leap in population growth.

Throughout the first half of the 20th century, chemistry was the driving scientific force, creating a myriad of new compounds which would be used for industry, agriculture and military purposes, and in homes and workplaces throughout the world.

A second group of chemicals which have ultimately had a major effect on human life, health and survival have been the biocidal agents, including pesticides and herbicides. These substances have been the 'magic bullets' against agents such as insects and weeds which harm or destroy crops or are the transmitters of diseases to humans and animals. Many of these compounds are synthetic chemicals: in other words, they do not occur naturally in nature, but have been synthesized by man. The introduction, in the latter days of World War II, of DDT to combat the mosquito-borne disease malaria is a classic case of pesticide use (see Box 1.1). Following World War II, agricultural practice

---

Box 1.1: The DDT story

Dichlorodiphenyltrichloroethane, more commonly known as DDT, became a household name as a pesticide and a saviour against a wide range of insect pests. Although it was not introduced until the latter days of World War II, DDT was first synthesized by the German chemist Othmar Zeidler in 1874. However, its power as a pesticide was not recognized until 1939, after which its use became widespread.

The first widespread use of DDT was by the military during World War II. Diseases such as malaria and dengue fever could thus be controlled simply by the application of the 'magic dust'. DDT has been credited with controlling typhus in Naples during 1943 and the eradication of malaria in Italy and

continued

Box 1.1 *continued*

Sardinia during 1945, and was widely used in the Pacific theatres of war to control insect pests. The advantages of DDT seemed obvious: it appeared to be nontoxic to humans who liberally dusted it on their skin and clothing; it was long lasting when applied to the environment; and above all it was effective. But the side-effects were soon to be seen in certain quarters.

DDT was used extremely effectively in the control of malaria throughout Asia by the World Health Organization in the period following World War II. But, in Malaysia, an interesting phenomenon occurred. The thatched roofs of villages where DDT had been sprayed collapsed, and the local cats died. How could this be so?

We now know the reasons. Apart from the vector of malaria (a mosquito), DDT also killed a wide range of other insects. These included a wasp that usually fed upon moth larvae in the thatched roofs of the villages. The wasps were more sensitive to DDT than the moth larvae, so the wasps died and, with the lack of a natural predator, the larvae flourished. The larvae fed upon the thatch, and the roofs collapsed.

By the 1960s, DDT use was widespread in agricultural practice, and an enormous amount of the substance was used annually to control insect pests. Birds eating insects accumulated large amounts of DDT in their bodies, and died or failed to reproduce as a result. The household use of DDT added to this; DDT was an effective mothproofer within the home, and an effective pesticide in the garden. Although DDT was undoubtedly a success in controlling such diseases as malaria, its widespread and uncontrolled use was increasingly seen as a problem. Ecotoxicological studies detailed its presence in populations of living organisms worldwide. DDT was discovered in living organisms from the Arctic to the Antarctic: it was suddenly a problem on a global basis. The problem was exacerbated by the realization that DDT caused eggshell thinning in many species of birds, leading to ineffective hatching and loss of populations. The ecotoxicological writing was on the wall.

In 1971, the USA banned the use of DDT for most purposes. It was replaced with a variety of other pesticides, but the damage was done. Although DDT was effective, its overuse had led to global environmental contamination. Its era of massive use was over.

If nothing else, the DDT story is a case history which illustrates the dangers of indiscriminate usage of substances. In the same manner that many antibiotics have become ineffective against disease, indiscriminate use of DDT has caused global problems beyond our initial imagination. The sheer fact that DDT was a persistent chemical in the environment—i.e. it did not degrade rapidly—was an added factor in the chain of events which has led us to realize that use of such compounds needs to be strictly controlled, and that effectiveness over long periods alone is not a formula for the continued health of the environment. DDT certainly sparked the 'Age of Pesticide Use', but the end result has been a sobering experience.

became almost as dependent upon biocidal agents as it was on the use of fertilizers. An enormous range of new compounds was introduced, often when previously used compounds became ineffective due to the development of resistance amongst target organisms. Many compounds were designed not only to kill organisms, but to stay active in the environment for extended periods, thus lowering the costs involved in their usage. Still other compounds were designed to be 'broad spectrum'—killing more than one pest at a time (often, unfortunately, including 'nontarget' species).

The period since the end of World War II has seen a rapid development in technology, communications and transport, a vast widening of the usage of various chemicals and a huge demand for the raw materials from which they are derived. The use of various metals for technology has also increased substantially, and previously rarely utilized metals, including mercury and cadmium, have been used in increasing quantities, often with toxicologically significant side-effects for the environment and for human populations. Mercury is an excellent case in point (see Box 1.2).

Box 1.2: The mercury story

Minamata disease has become a catch-phrase in the annals of trace metal contamination. The disease was first recognized in Japan, in the vicinity of Minamata Bay, in the mid-1950s.

In this area, a factory manufacturing acetaldehyde discharged its wastes into the Bay. The wastes included mercury in various forms. Certainly, some methyl mercury was discharged in the waste stream, although this form of the metal can also be formed from the element mercury itself in the environment through the action of microorganisms. Huge amounts of mercury were discharged into Minamata Bay—estimated at some 600 tonnes between 1932 and 1970.

Within the Bay, methyl mercury was accumulated by living organisms. This form of mercury is lipophilic (i.e. fat soluble), and easily accumulated from the water column by the fatty tissues of living organisms. The early problems noticed in the area related to the local organisms—birds failed to fly in the correct manner, and various reports had them dropping from the sky into the sea for unknown reasons. Some organisms within Minamata Bay died, whilst others suffered changes in behaviour, leading local children to be able to catch octopus with their hands. In addition, other unusual phenomena were reported: cats died, often in convulsions.

During the 1960s, the local people also became ill. The disease appeared to affect the central nervous system, resulting in convulsions, excessive salivation and staggering. Some people, including newly born children, died—often in horrible circumstances.

continued

Box 1.2 *continued*

The cause of these strange events was eventually tracked to the mercury discharges into Minamata Bay, and more specifically to the accumulation of methyl mercury within the bodies of people eating food from the Bay. As an interesting side-effect, it was seen that pregnant women suffered less than others, although their newborn children suffered from neurological disorders, such as palsy and retardation, since some of the mercury was passed onto these children.

The phenomenon of bioaccumulation was producing these effects. Methyl mercury passed through the local food chains, accumulating in high concentrations in fish, birds and the organisms which fed upon them—cats and humans. Unlike the DDT case (see Box 1.1), mercury had entered the environment through a deliberate discharge of industrial waste. But the overall effects were the same—bioaccumulation, leading ultimately to the death of living organisms.

Apart from biocidal agents and various metals, many new compounds have been extensively used for various industrial purposes. For instance, the polychlorinated biphenyls (PCBs), which were used principally as dielectric (insulating) fluids in the electrical industry, were first introduced in 1929. The chlorofluorocarbons (CFCs), so widely used in refrigeration and air-conditioning, were introduced just shortly after. In addition, plastics of various kinds have been enormously used by man. It was only many years after the introduction of these compounds that environmental problems caused by their use and careless disposal were first recognized. Since the 1960s, many other substances have been similarly acknowledged as environmental pollutants. It was in 1962 that Rachael Carson published her influential book *Silent Spring*, detailing to the general public for the first time the unintended effects of widespread chemical usage (especially pesticides) on wildlife.

Waste and its sources

Rachael Carson's book, *Silent Spring*, proved a revelation and in the next two decades, a myriad of chemicals were revealed to be present in the environment at what were possibly toxic levels. The scientific search for these chemicals revealed that pesticides, herbicides and a wide variety of industrial chemicals (including the trace metals and such substances as PCBs) were almost ubiquitously distributed throughout the globe.

A number of factors were highlighted by this search. Firstly, poor *waste management* and poor *waste disposal practices* were the key factors in the spread of potentially toxic compounds in the environment. Secondly, *deliberate* and *unintentional* releases of various compounds were differentiated as sources.

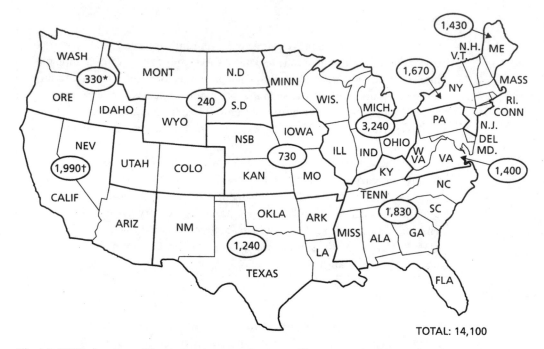

Fig. 1.2 US Environmental Protection Agency estimates and locations of facilities producing hazardous wastes in 1981.

Waste disposal practices in the early part of the 20th century were mostly at a very basic level. Although a number of large cities had sewage treatment facilities, many areas still resorted to practices of disposal into rivers, lakes or oceans, onto the land, or into pits dug in the ground, with very little (if any) treatment prior to disposal.

However, the realization that toxic wastes could contaminate the environment changed these practices, and governments increasingly legislated for enhanced and vastly improved disposal practices. These became increasingly sophisticated. Sewage disposal methods needed to eliminate potentially pathogenic bacteria and excessive nutrients from entering and contaminating the environment; rubbish disposal practices were required to take into account the need to separate waste streams, in order to extract materials that could be recycled and reused, as well as to dispose safely of materials that were nonreusable or potentially toxic.

In the early 1980s, the US Environmental Protection Agency (USEPA) undertook wide-ranging surveys to estimate the number of facilities producing toxic wastes (see Fig. 1.2) and the sites of the worst toxic waste dumps were identified as those places where potentially dangerous materials had been dumped in quantities deemed to be potentially harmful.

Landfilling—the practice of placing waste materials in a pit dug in the soil and covering it with earth—was, and still remains one of the key methods for the disposal of wastes, including toxic wastes. But a number of problems have been experienced with landfilling. Amongst the most notorious cases is that of Love Canal in Niagara Falls, New York, USA, where wastes in a landfill leached out into the basements of houses built later on the site, not only causing neurological and reproductive problems, but also being implicated in carcinogenesis (i.e. the causation of cancer). Cases such as this have resulted in a considerable tightening of legislation related to the disposal of wastes in landfills, including measures for the separation of wastes prior to disposal (to prevent chemical reactions occurring after burial), and addressing the need for adequate design and operational criteria for the dump sites themselves.

Concomitant with these changes have been advances in alternative methods for the disposal of wastes. Thus, methods such as high-temperature incineration, deep-well injection and immobilization of toxic wastes have been put forward. However, it must always be remembered that no one method is infallible. For instance, incineration can result in the formation of compounds more toxic than the original waste. Such is the case when PCBs and organochlorine pesticides are burnt using inefficient high-temperature incinerators—the end result is the formation of dioxins as a by-product, compounds considered to be among the most toxic of waste compounds.

Waste disposal practices also highlight the second key factor in the spread of toxic waste materials—the concepts of point and nonpoint sources, and deliberate and unintentional discharges. *Point sources* are easy to recognize: a point source is simply a site recognizable as the source of wastes (and potentially toxic materials) into the environment. Thus, a sewage outfall to a river or lake is a point source; similarly, a factory pipeline carrying liquid wastes to the coastal environment is also a point source. By definition, they are also *deliberate* sites of disposal.

Nonpoint sources are diverse, and are not easily defined in terms of a certain outlet, pipeline or source. Often, they are the result of *unintentional* disposal of wastes, or perhaps a deliberate usage has resulted in an environmental input which cannot, for one reason or another, be subsequently defined as an actual source. Rather, nonpoint sources refer to potentially toxic material which enters the environment from a variety of different inputs. For instance, in the city situation, pollutants may enter rivers and coastal waters through run-off from storm water. The storm water which enters the sea has come from all over the city, and contains contaminants from roads (e.g. petroleum hydrocarbons), gardens and parks (e.g. fertilizers and pesticides) and factories and commercial areas (e.g. various industrial chemicals, including such substances as PCBs). In such cases, the actual source of contaminants cannot be definitely pinpointed.

Development of the ecotoxicology concept

Living organisms are composed of cells, inside which complex biochemical reactions take place which form the basis of life. In the more complicated

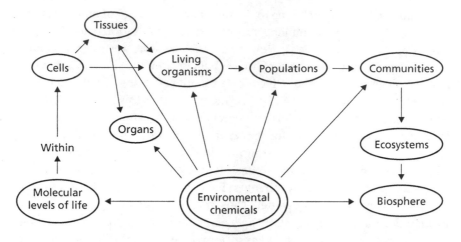

Fig. 1.3 Levels of biological organization and their interactions with environmental chemicals.

plants and animals, similar cells may be grouped together as tissues, and tissues become grouped as organs, which are committed to undertaking certain specific functions. Living organisms and the cells, tissues and organs of which they are composed require food and essential nutrients in order to grow and replicate. These materials are obtained, of course, from the environment, or by feeding on other living organisms.

Plants and animals need to reproduce, so similar species are found close to each other, forming populations. Within certain areas, populations interact with other species, forming local communities which are often based around certain physical, geographical or meteorological features (e.g. rocky shore communities; coral reef communities, etc.). At the next level of organization, communities interact with each other and their wider environment to form ecosystems, which can be defined as regional communities, which of course also interact with their physical environments (e.g. rain forest ecosystems; tundra; deserts, etc.). At the highest level of organization—the biosphere—we can consider all the organisms on the earth as a vast, global ecosystem.

Thus, life operates at a range of levels: at the molecular or subcellular level within cells; at the cellular level itself and within tissues and organs; at the level of the individual organism; within populations and communities; and ultimately within ecosystems and the biosphere as illustrated in Fig. 1.3.

The major concerns of ecotoxicology are the interactions between living organisms and toxic chemicals in the environment. We can thus view ecotoxicology at various levels, and the various relationships are illustrated in Fig. 1.4. This figure will be a very important component of our future discussions, and will form the basis for the structure and content of this book.

If ecotoxicology concerns toxic substances in the environment, chemistry has a major role to play. Chemistry elucidates the nature of the chemicals

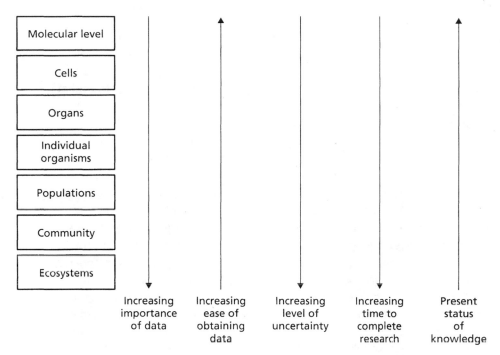

Fig. 1.4 Relationships of aspects of the science of ecotoxicology and the different levels of biological organization.

involved, and through judicious analysis of environmental samples, can aid in pinpointing the *source of contaminants*. Much information about these sources can also be gained from examining the nature of waste material being added to the environment—for example, different types of industry produce different types of wastes, and often waste materials are typical of a particular type of industry.

Chemical analysis can also provide information on the *environmental distribution* of chemicals. This requires analysis of various environmental phases so that pollutant loads in each can be estimated. *Pollutant load* information provides us with details of the bulk (or total amount) of particular pollutants entering or present within the environment, and which compartments of the environment may be at greatest risk from their effects. Thus, analysis of air, water, soil and living organisms may be required. Chemical analyses of living organisms are most important in this regard, as these data tell us how *available* a particular pollutant is to living organisms.

Not all pollutants are *biologically available*, and some pollutants accumulate in living organisms to a greater extent than others. For instance, some pollutants attach strongly onto soil and sediment particles and become effectively 'unavailable' to living organisms. Others may be biologically available, but are rapidly transformed or broken down within living organisms. Still others may be transformed in the environment, either through abiotic transformations (e.g. by processes such as phototransformation) or by microorganisms such as bacteria and fungi. This transformation process is of considerable importance, as it can alter the risks that certain pollutants pose. Some pollutants, once transformed, may become totally benign compounds through either partial or complete degradation; in other cases, compounds even more toxic than that which was originally released may be formed. Such is the case with mercury, for example (see Box 1.2).

The regular analysis of environmental samples is commonly known as *monitoring*. Chemical monitoring of the environment has been the mainstay of ecotoxicological studies for many years, and most government agencies have a wealth of historical data detailing concentrations of various chemicals in air, water, soils, sediments and biota. In the early years of ecotoxicological studies —during the 1960s and 1970s in particular—these chemical studies were heavily relied on to provide spatial and temporal data on contaminant distribution, and, as well as being believable, the data they produced also rapidly assumed a major role as the basis of environmental protection procedures.

But of course, living organisms remove contaminants from their environment, and may sequester them within their bodies—a phenomenon called *biological uptake*. Although we can measure the amounts of these contaminants in living organisms through chemical techniques, this tells us little about the way in which the *organism responds* to the chemicals. This response is dependent upon the nature of the chemical and the actual dose received by the organism. This dose and corresponding response equation has been recognized for centuries—and is the basis of the quotation from the works of the 16th century physician Paracelsus;

All substances are poisons. There is none which is not a poison. The right dose differentiates a poison from a remedy.

Measurement of the response to environmental chemical exposure has become a rapidly expanding branch of ecotoxicology. Many different techniques can be used for this purpose: measurements of changes in enzyme activity and changes in cell biochemistry and structure; measuring alterations to behaviour or changes in an individual's reproductive capability; and measuring various alterations to basic physiological activities such as respiration, photosynthesis, locomotion, feeding behaviour, etc.

Measuring responses at the level of the individual organism is often referred to as *biomonitoring*. Measurement of chemicals within the tissues of an organism may also be considered as biomonitoring. For example, this technique has formed the basis of the 'Musselwatch' programme in the USA, which uses bivalve shellfish (especially mussels) as indicators of contaminants in estuarine and marine waters. Mussels are ideal for this purpose: they filter feed and during this process accumulate certain contaminants within their tissues. This has a great advantage over monitoring contaminants in water, as many contaminants are present in waters at very low (but nonetheless toxicologically significant) concentrations. Thus, mussels act as *concentrators* or *bioaccumulators* of chemical contamination, resulting in analysis becoming an easier task for the chemist. Mussels have been found to be very useful for indicating trace metal pollution, as well as contamination caused by certain organic compounds such as the pesticides and PCBs. These pollutants are often collectively referred to as *conservative* contaminants, as they do not change rapidly in the environment and their essential nature is thus conserved.

Of course, the measurement of a huge variety of other responses by living organisms upon exposure to chemicals is possible in biomonitoring. The response measured is often referred to as the *endpoint* of a toxicological test. Ecotoxicologists seek to measure these endpoints in order to establish a relationship between the concentration of chemicals in the environment, and the response they elicit in living organisms.

Specific responses by living organisms can act as indicators or *markers* of changes elicited as a result of exposure to a chemical insult. The responses are often referred to as *biomarkers,* and their measurement has become increasingly important in ecotoxicology. Biomarkers illustrate for us the importance of the dose–response relationship in ecotoxicological studies. Earlier work in medicine and pharmacology (e.g. the development of new drugs) relied heavily on such a concept, and this has been adopted into the environmental context.

In addition to providing subcellular and cellular responses, individual organisms can often form the basis of toxicity tests. The most frequently used endpoint of these tests is death of the organisms. Such tests are able to measure the dose that causes death (the result of *acute exposure* to a chemical insult). Responses which indicate the organism may be, or is about to be, suffering from stress due to a chronic exposure are also exploited. These tests may be conducted with single chemicals or with mixtures. The latter, of course, is the more likely situation in the environment, and responses from such tests can be extremely useful in environmental protection and management.

When individual organisms respond to the presence of chemicals in the environment, it is conceivable that, sooner or later, populations should also respond. As effects ripple through the hierarchy of organization from sources of pollutants to individual organism response then effects should also be visible at the community and ecosystem levels in the course of time as shown in Fig. 1.5. Many biologists undertake ecotoxicological investigations at these different levels of organization. The accumulated data can be referred to as *ecological data*, and this has arguably been the most frequently investigated area in past ecotoxicological studies. After all, this is the 'most visible' level of biological organization, and is the area in which the general public—and politicians —take the most interest. Ecosystems are the levels that demand protection, and around which most legislation is written. Thus, ecologists have collected information on the changes which occur to populations, and how species diversity and other relevant parameters change within communities as a result of exposure to chemicals.

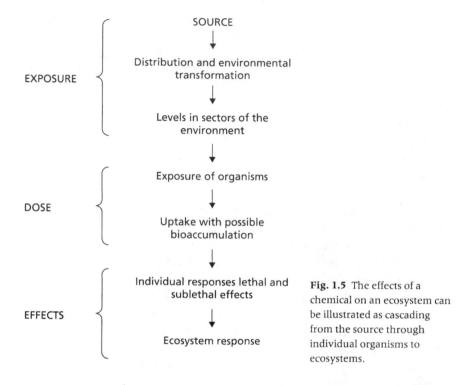

Fig. 1.5 The effects of a chemical on an ecosystem can be illustrated as cascading from the source through individual organisms to ecosystems.

Biological measurements of the impact of toxicants is much easier—and more accurately performed—at lower levels of organization. Thus, biomarker measurements, or measurements at the level of individual organisms, can be relatively easily achieved. But do these reflect, in turn, likely changes at higher levels of organization? The big challenge of ecotoxicology is to relate the more easily achieved measures at the biomarker level to effects that occur at the community or ecosystem level, where measurements are much more difficult to interpret, are often confused with natural events, and take much longer to perform. (The levels of biological organization related to ease of performing toxicity tests and the importance of the data to environmental management are illustrated in Fig. 1.5.) Ecotoxicology strives for a 'fast response' to chemicals in the environment—the ability to rapidly predict events that may occur in the future. Measurements at higher levels of organization, by their very nature, are slow to obtain and difficult to interpret.

In modern ecotoxicological research, chemistry needs to be closely integrated with biological studies. Biological studies also need to be undertaken at multiple levels of organization—from the subcellular to the ecosystem level. The integration of these data provides the important element of *risk assessment* in the final ecotoxicological equation: evaluating what compounds are present, their possible effects and the likelihood of their causing major damage.

This is, of course, no easy task, but it is a necessary one, and one which environmental managers, legislators and governments have increasingly had to come to grips with.

Retrospective ecotoxicology

Retrospective ecotoxicology refers to the use of previously established data in the understanding of a toxic response to chemicals in the environment. In essence, retrospective ecotoxicology is a form of 'environmental epidemiology'.

Within medicine, epidemiology is the scientific branch which seeks to establish the causes of various diseases and the manner in which they spread through populations. One of the earliest successful examples of epidemiological research involved a cholera epidemic in London in 1849, during which a local doctor, John Snow, realized that a common cause was associated with the victims: they all obtained their drinking water from a single site (the Broad Street Pump). His consequent action—simply removing the handle of the pump—brought the epidemic to a close. What Dr Snow had done was to recognize cause and effect through a retrospective study of those who contracted the disease.

Retrospective studies have also been employed in the study of cancer-causing agents in the environment. Such studies attempt to correlate the distribution and usage patterns of chemicals with the incidence of cancer in exposed (or potentially exposed) populations. Epidemiological studies have often been viewed as the most reliable means of identifying carcinogenic compounds, especially in human populations. But, by their very nature, such studies always take place after exposure has occurred. In the case of carcinogenic compounds, 20 years or more may have passed from the time of exposure to the first sign of symptoms. This means of course that retrospective studies are not useful in the *prediction* of carcinogenic effects of new compounds used in industry or introduced into the environment.

Epidemiological studies of carcinogenic chemicals have other problems as well. They are poor at identifying chemical agents which result in very rare forms of cancer. In addition, it is often difficult to discern the actual cause of carcinogenesis from a myriad of different factors—including other chemicals, differences in exposure and diet, the influence of smoking (amongst human populations) and genetic predisposition, to name but a few. Retrospective epidemiological studies of this type are also expensive to conduct, and establishing a suitable control population against which the 'exposed' population can be compared is very difficult indeed. Notwithstanding these criticisms, retrospective ecotoxicological studies allow us to recognize that a situation has occurred; to take remedial action; and to set guidelines to ensure that in the future the environment is protected against similar occurrences. However, retrospective ecotoxicological studies are only one side of the coin. Predictive ecotoxicology has become an arguably more important tool, which allows us to establish the possible fate and effects of chemicals which may be introduced into the environment in future years.

Predictive ecotoxicology

Predictive ecotoxicology, as its name suggests, seeks to *predict* the effects of chemicals that may reach the environment. In this instance, it is hoped that through adequate research and testing techniques enough will be known about the effects of a particular chemical to be able to predict its likely effects. So, given the huge number of new chemicals introduced into industry each year, what is needed in order to be able to undertake this very important task of seeing into the future?

A number of different approaches are involved in predictive ecotoxicology. At the front line is a *theoretical* approach to the problems involved. Modern chemical techniques are able to predict the behaviour and activity of various chemicals, based upon their structure and their known properties. Thus, for example, the likelihood of a chemical being accumulated by living organisms and sequestered in their tissues for extended periods can be predicted based on this information. Chemical structure can also be used to predict the likelihood of toxic actions.

This information is often the basis for the prior assessment of chemicals before they are used in industrial or domestic situations. For instance, newly developed chemicals are subject to a battery of predictions and tests prior to their release, and theoretical aspects of their toxicology are just one of the tools used. In addition, a wide variety of toxicity tests can also be employed to predict whether a compound or a mixture of substances (for example, the complex effluent from an industrial process) is likely to cause environmental effects. Such toxicity tests are practical (experimental) approaches to ecotoxicology, and involve the sorts of tests that have been previously described—for example, tests involving biomarkers of various sorts, combined with acute and chronic toxicity tests on whole organisms.

The predictive approach to ecotoxicology allows new chemicals to be screened before they are used, and also allows industry to predict the effects of their effluents and wastes before they are released to the environment. In combination with the theoretical and practical approaches used in this process, many governments have instituted legislation that ensures compliance before the introduction of a new chemical, as well as during the period of its use. Such legislation insists that adequate tests be performed on new chemicals prior to their introduction. Producers of various wastes and industrial or municipal effluents are also required to test materials before they are released into the environment in waste streams. An example of this *legislative* approach to the control of contaminants occurs in the state of California, USA, where an 'Ocean Plan' to limit or prevent toxic wastes from entering into marine and estuarine environments has been instituted. This legislation, amongst other requirements, insists on the use of toxicity tests, using a variety of different living organisms from different trophic levels to assess wastes both prior to disposal and within receiving waters where they may be diluted but may be still present in toxicologically significant concentrations.

Predictive ecotoxicology is the mainstay of modern ecotoxicological monitoring and a keystone of modern environmental protection methods. The concepts involved rely on the dose–response relationship which we have already seen to be a necessary part of establishing toxic responses of chemicals. But equally, chemistry plays a role: routine monitoring of environmental compartments is also necessary to establish that undesirable concentrations are not occurring. In order to protect environments at the level of populations, communities and ecosystems, ecological monitoring must also be undertaken.

Thus, modern ecotoxicology is a complex of procedures, and involves measures and tests conducted at all levels of organization: from chemical assessments in various environmental phases; through biomarkers and toxicity tests; to ecological assessments at higher levels of organization. Ultimately, laws made by governments, at local, regional and federal levels, interact with science to provide a complex system of protection for the environment.

Conclusions

The 'keys' to ecotoxicology are based on a series of basic concepts, which lock into place the science of understanding how our environment interacts with toxic chemicals, and what we should do to protect harmful effects from occurring. These keys include:
• a study of chemistry, which tells us what types of chemicals are found in the environment, the quantity that is present, and their distribution in key environmental phases such as air, water, soil and biota;
• a study of toxicology, which elucidates the dose–response relationship between chemicals in the environment and the organisms with which they interact;
• a study of ecology, which demonstrates how populations, communities and ecosystems respond to toxic contaminants; and
• the legislative aspects of ecotoxicology, which seek to ensure that adequate protection is provided by law for all environments.

Uniting these aspects has resulted in the multidisciplinary science of ecotoxicology. More detailed aspects of this science will be expounded in the following chapters of this book. The diagram which formed a key part of this chapter, Figs 1.4 and 1.5, will be used throughout the book to demonstrate how various areas of ecotoxicological study fit in with others. We shall start with chemical aspects by examining the types, properties and sources of various contaminants. The chemical theme continues, with the distribution and transformation of chemicals in the environment as the central theme.

In later chapters the focus shifts towards the biological responses of contaminants. Molecular, biochemical, physiological and behavioural responses of organisms are highlighted, including mutagenesis, carcinogenesis and ter-atogenesis. Later the higher levels of biological organization are examined: in other words, the possible responses of populations, communities and ecosys-tems to toxic chemicals. The important relationship between the dose of con-taminants present in the environment, and the response produced as a result in living organisms is considered. Biomarkers and biomonitoring of hazards in the environment are considered. Risk assessment links most of the approaches and concepts presented previously.

The final chapter, Chapter 9, introduces the policy and management aspects, providing an overview of the ecotoxicology of chemicals and the importance of evaluating of the ecotoxicology of *commercial* chemicals in particular.

Further reading

Carson, Rachael (1962) *Silent Spring*. Fawcett Books, New York.

Forbes, V.E. & Forbes, T.L. (1994) *Ecotoxicology in Theory and Practice*. Chapman & Hall, London.

Francis, B.M. (1994) *Toxic Substances in the Environment*. John Wiley & Sons, New York.

Moriaty, F. (1988) *Ecotoxicology. The Study of Pollutants in Ecosystems*, 2nd edn. Academic Press, London.

Newman, M.C. (1995) *Quantitative Methods in Aquatic Ecotoxicology*. Lewis Publishers, Boca Raton, USA.

Rombke, J. & Moltmann, J.F. (1995) *Applied Ecotoxicology*. Lewis Publishers, Boca Raton, USA.

Walker, C.H., Hopkin, S.P., Sibly, R.M. & Peakall, D.B. (1996) *Principles of Ecotoxicology*. Taylor & Francis, London.

Zakrzewski, S.F. (1997) *Principles of Environmental Toxicology*, 2nd edn. American Chemical Society Monograph 190, American Chemical Society, Washington, DC.

2: Sources, Types and Properties of Toxicants

Introduction

Toxicology is one of the oldest sciences; it dates from ancient times starting with the search for remedies and cures for ills. Despite this long history there can be difficulty in defining which substances are in fact toxic, and those which are not. Paracelsus is widely quoted from his classic book published in the 1500s as saying that all substances are poisons, only the dose makes a distinction between one which is a poison and one which is a remedy. This statement remains very true today and emphasizes the importance of establishing the level of exposure to a toxicant in order to evaluate toxic effects. In this chapter *toxicants* are considered to be substances which are toxic in relatively small doses and do not originate from animals and plants and thus are not natural agents. Toxic substances of natural biological origin, principally derived from microbes, plants and animals, are usually described as *toxins*. There are a few substances of natural origin which can be important *ecotoxicants*. For example, the toxins from Cyanobacteria or blue-green algae may be important in bloom situations and may influence the natural ecosystem. Although these toxins are of natural origin they often occur as a result of a human-induced situation. For example, elevated concentrations of plant nutrients, resulting from sewage and other discharges, can stimulate excessive growths of blue-green algae.

The broad range of toxicants which are relatively common in our society includes petroleum, pharmaceuticals and pesticides. But there are substances within this broad group of toxicants which are discharged into the environment and thus have a specific relationship to natural ecosystems, since natural ecosystems are exposed to them. These can be referred to as *ecotoxicants* and are defined primarily by their usage which leads to distribution in the natural environment and to the exposure of natural ecosystems to these substances. These substances could be discharges from industry such as petroleum hydrocarbons, heavy metals, acids, alkalis, solvents and so on. Domestic activities can also result in discharges which cause environmental contamination. Common contaminants from domestic activities are substances such as pesticides used in the garden, polycyclic aromatic hydrocarbons from motor vehicles and so on. Agricultural activities can also give rise to a range of ecotoxicants, principally pesticides. It should be noted that not all chemical substances in discharges can be considered to be toxicants. In particular some of the activities mentioned above result in discharges of nitrogen and phosphorus compounds contained in low concentrations in domestic sewerage and agricultural fertilizers. These are not considered to be ecotoxicants since these substances are not toxic in relatively low concentrations. The principal effects of nitrogen and phosphorus as plant nutrients include increased primary production but there can be secondary toxic effects at a relatively low level resulting from these substances in some situations.

The term ecotoxicant encompasses a broad range of substances since this group includes all substances which are discharged into the environment and have a potential impact on ecosystems in relatively low concentrations. It can include substances which are already present in the environment but also occur in discharges, adding to the environmental occurrence. For example, many metals and related elements occur naturally in the environment in significant concentrations. Lead, arsenic, mercury and many other substances occur in the oceans, soils and other parts of the environment. Polycyclic aromatic hydrocarbons produced by combustion are present in low concentrations as a natural product of combustion in the environment but levels have been increased by the expanded usage of fuels leading to discharge of elevated concentrations to the environment. A somewhat similar situation occurs with petroleum hydrocarbons. Thus for many substances which can be considered to be ecotoxicants, there already exist background concentrations of substances within the environment in general and elevated concentrations in specific sectors.

In addition to the ecotoxicants mentioned above there are a wide range of synthetic compounds distributed or used in the environment for a variety of purposes. For example, chemicals such as dichlorodiphenyltrichloroethane (DDT) and other chlorohydrocarbons, glyphosate, 2,4-dichlorophenoxyacetic acid (2,4-D) and so on are totally new to the environment since they are produced by synthetic chemical processes in industry. Industrial chemicals such as the polychlorobiphenyls and most solvents are used and can be distributed accidentally, or after use, into the environment. The organometallic compounds, such as organotin compounds (e.g. tributyltin), which are applied to the hulls of ships as an antifouling compound can be distributed into the environment as well. These compounds may leach or diffuse into the aquatic environment and result in damage to the aquatic ecosystem.

The origin of an ecotoxicant, whether natural or synthetic, makes little difference to its potential ecological effects. For example, some highly toxic substances, such as mercury and cadmium, are natural components of the environment and occur in seawater in all locations and in soils in specific locations in relatively high concentrations. On the other hand, there are common synthetic chemicals which are discharged into the environment with little ecological effect. A good example of a synthetic compound in this group is soap, the salt, usually sodium or potassium, of long-chain fatty acids. This substance is not a natural component of the environment but is produced by chemical synthesis using fat and alkali and is distributed into the environment in large amounts from human society. However, ecological effects observed as a result of this large scale discharge have been relatively minor. Thus there is no general rule for the nature, origin and effects of ecotoxicants, and each compound must be judged on its properties and its observed effects.

Sources of ecotoxicants

Some of the potential sources of ecotoxicants are shown in Table 2.1 together with the natural environments which could be affected. It can be seen that most activities of human society produce toxicants. Sewerage is often considered in terms of its biochemical oxygen demand (BOD) and ability to reduce the dissolved oxygen in receiving water as well as the presence of nitrogen and phosphorus which can cause eutrophication in aquatic areas. But in addition to these substances sewerage usually contains a range of ecotoxicants at low concentrations. The volume of sewerage is relatively high and so the total amount of toxicant discharged into the environment can be relatively large. For example, sewerage is a major source of petroleum hydrocarbons discharged into the oceans in comparison with discharges due to the operations of oil fields, spillages and so on. These trace toxicants in sewage mainly originate from industries discharging into the sewerage system. Similarly, stormwater also contains a wide array of ecotoxicants in low concentrations.

Table 2.1 Sources and types of toxicants discharged and environments affected.

Source	Some chemical groups involved	Environments affected
Motor vehicle exhausts, electricity generation and industrial discharges to the atmosphere	Lead and other toxic metals, carbon monoxide, carbon dioxide, aromatic hydrocarbons, sulphur dioxide, hydrocarbons, PCDDs, PCDFs, PCBs*	Human and natural terrestrial systems
Sewage	Aromatic hydrocarbons, hydrocarbons, chlorohydrocarbons, toxic metals, surfactants	Aquatic systems
Stormwater runoff	Aromatic hydrocarbons, hydrocarbons, lead and other toxic metals	Aquatic systems
Industrial discharges to waterways	Acids, toxic metals, salts, hydrocarbons, PCDDs and PCDFs*	Aquatic systems
Urban and industrial discharges to soil	Toxic metals, salts, hydrocarbons, PCDDs, PCDFs, PCBs*	Human and natural terrestrial systems
Rural industries	Chlorohydrocarbons, organophosphorus compounds	Human, natural terrestrial and aquatic systems

* PCBs, polychlorobiphenyls; PCDDs, polychlorodibenzodioxins; PCDFs, polychlorodibenzofurans.

These substances originate principally from discharges from motor vehicles which are deposited on the road surface and subsequently swept into stormwater. These toxicants are in waste waters and so have the capacity to have adverse effects on aquatic ecosystems in the receiving water.

Motor vehicles are a major source of lead and other metals, polycyclic aromatic hydrocarbons and toxic gases. All these substances are discharged into the atmosphere but many of the toxic compounds are in particulate form and can deposit into soils. So soils in urban areas can contain significant amounts of lead, polycyclic aromatic hydrocarbons and other ecotoxicants. Often, industrial wastes are disposed of directly into pits dug into soil and so the soil can be contaminated and possibly the ground water as well.

Outside urban areas agricultural activities are probably the major source of contaminants. However, mining activities can also release metals and other substances principally into waterways although the atmosphere and soil can also be contaminated. The growing of crops often involves the widespread use of pesticides including insecticides and herbicides which can contaminate soil. Stormwater runoff can then transport these toxicants into adjacent waterways.

Perhaps the most spectacular example of toxicant contamination of the environment occurs due to the accidental spillage of petroleum. Over the years many disasters of this kind have occurred releasing tens of thousands of tonnes of petroleum into the coastal environment. The immediate effects can be the deaths of many organisms but long-term sublethal effects may occur over a long period of time although the contaminant concentrations are reduced by environmental processes such as evaporation and oxidation.

Properties of ecotoxicants

The behaviour and ecological effects in relationship to the properties of an ecotoxicant are outlined in Fig. 2.1. The initial behaviour of a toxicant immediately after discharge involves physical dispersal in the environment which is influenced by physical state, i.e. gas, liquid or solid. The physicochemical properties principally govern the distribution of the substance into different environmental phases or components—air, water, soil, biota and so on. If a compound is water soluble, then it tends to favour aquatic phases such as the water in lakes, rivers, the sea and so on; whereas if it is lipid soluble it is more likely to be preferentially taken up by biota. Its biochemical properties influence the substance's ability to degrade or be transformed by biota, including microorganisms in the environment, and be excreted or bioaccumulated. Although most chemical processes are mediated by biota, some transformations can occur by abiotic mechanisms. After uptake by biota the distribution of the chemical within the organism may result in interaction with an active site leading to lethal or sublethal responses depending on the physiological properties of the substance. The reaction of the individual biota may modify

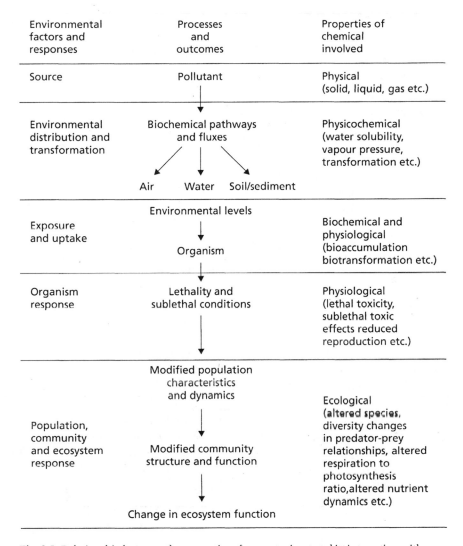

Environmental factors and responses	Processes and outcomes	Properties of chemical involved
Source	Pollutant	Physical (solid, liquid, gas etc.)
Environmental distribution and transformation	Biochemical pathways and fluxes Air Water Soil/sediment	Physicochemical (water solubility, vapour pressure, transformation etc.)
Exposure and uptake	Environmental levels Organism	Biochemical and physiological (bioaccumulation biotransformation etc.)
Organism response	Lethality and sublethal conditions	Physiological (lethal toxicity, sublethal toxic effects reduced reproduction etc.)
Population, community and ecosystem response	Modified population characteristics and dynamics Modified community structure and function Change in ecosystem function	Ecological (altered species, diversity changes in predator-prey relationships, altered respiration to photosynthesis ratio,altered nutrient dynamics etc.)

Fig. 2.1 Relationship between the properties of an ecotoxicant and its interaction with ecosystems.

the population of a species present perhaps because only the juveniles or some other segment of the population may be susceptible. These individual and population responses in turn may lead to certain species being removed from the system giving a modified community structure and finally a resultant change to the functioning of the ecosystem. For example, flows of energy and nutrients through the ecosystem may be altered. However, little is known about the ecological properties of chemicals, since most research has focused on the physical, physicochemical, biochemical and physiological properties of the substances. These properties can be measured relatively easily in the laboratory whereas ecological properties are difficult to measure in a standard way.

Category	Examples
Physicochemical properties	K_{OW} value Aqueous solubility Adsorption/desorption on soil
Degradation and accumulation	Biodegradability Fish bioaccumulation
Effects on biotic systems	Algae—growth inhibition Fish—acute toxicity Terrestrial plants—growth test
Effects on mammals	Oral toxicity—rats Eye toxicity—rabbits Reproduction toxicity—rats Carcinogenicity—rodents

Table 2.2 Types of laboratory-based tests used to evaluate adverse effects on ecosystems.

Some common examples of properties which can be measured in the laboratory are shown in Table 2.2. Most of the evaluations of the ecotoxicology of a chemical are based on extrapolations from the laboratory-based tests of the type shown in Table 2.2, which are described in Chapter 6.

When evaluating and considering the ecotoxicology of a chemical, properties which seem to be particularly important are often used to group and describe chemicals. So ecotoxicology can be used as a basis for positioning chemicals into groups that share similar properties. If we classify chemicals in this way the chemicals within the groups give a better idea of the properties of a single chemical than can be gained from considering the single chemical alone. In addition, if the substance can be placed within a category then many of its other properties are suggested by the properties of the group into which it is classified. Some of the common groupings for ecotoxicants are shown in Table 2.3. Classifications based on the observed biological effects are very common, so a substance can be described as an insecticide or a herbicide and so on. Groups based on chemical structure are also common and help in identifying the chemical properties which a compound within a particular group could be expected to possess. Common chemical classifications into groups are shown in Table 2.3. Groups based on important physicochemical properties are also commonly used, these groups giving an idea of the physicochemical behaviour of a compound within a particular group. For example, if a compound is classified as surfactant then a set of properties associated with this group are indicated for a compound in this category. These groups are not mutually exclusive but overlap considerably. In fact often combinations of terms from several groups are used to describe a chemical. For example DDT can be described as a pesticide, a chlorohydrocarbon pesticide or a lipophilic chlorohydrocarbon pesticide.

Table 2.3 Common terms for and characteristics of groups of ecotoxicants.

Term	Characteristics	Examples
Biological property-based group		
Pesticides	Toxic to pests	DDT, 2,4-D
Insecticides	Toxic to insects	DDT, parathion
Herbicides	Toxic to plants	2,4-D, glyphosate
Fungicides	Toxic to fungi	Phenyl mercury acetate
Rodenticides	Toxic to rodents	Hydrogen cyanide
Carcinogens	Induce cancer	Benzo[a]pyrene
Chemical structure-based group		
Chlorohydrocarbons (chlorinated hydrocarbons)	Compounds based on chlorine, carbon and hydrogen alone	DDT
Hydrocarbons	Compounds based on carbon and hydrogen alone	Hexane
Polycyclic aromatic hydrocarbons (PAHs)	Polycyclic aromatic hydrocarbons containing two or more aromatic rings	Benzo[a]pyrene
Organochlorines (OCs)	Organic compounds containing chlorine	DDT, 2,4-D
Persistent organochlorines	Organic compounds containing chlorine with environmental persistence	DDT
Dioxins	Combustion and industrial products having polychlorodibenzodioxin structure	Tetrachlorodibenzodioxin
Furans	Combustion and industrial products having polychlorodibenzofuran structure	Tetrachlorodibenzofuran
Heavy metals	Toxic metals of higher specific gravity	Mercury
Organometallics	Organic compounds containing metals	Tributyltin
Pyrethroids	Usually synthetic pesticides related to pyrethrum	Fenvalerate
Organophosphates	Compounds based on organophosphate structure	Parathion
Phenoxy acetic acids	Compounds based on phenoxy acetic acid structure	2,4-D
Physicochemical property-based group		
Lipophilic (or hydrophobic)	Fat soluble and water insoluble	DDT
Hydrophilic	Water soluble and fat insoluble	Phenol
Neutral organic compounds	Organic compounds without ionic charges	DDT
Radionuclides	Substances having radioactivity	Uranium
Surfactants	Compounds which act at interfaces by altering surface tension	Alkylbenzene sulphonates

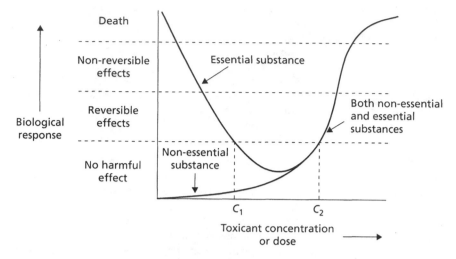

Fig. 2.2 Biological responses to concentrations of substances which are essential and nonessential to growth. Normal metabolic activity occurs between concentrations C_1 and C_2.

General effects of a toxicant on ecosystems

With organisms of a particular species a generalized perspective of responses is shown in Fig. 2.2. For substances which are essential for growth there can be no growth if the concentration is zero so death results. As the concentration of the essential substance increases so the severity of the adverse response declines until normal activity occurs between a concentration of C_1 and C_2. However, as the concentration increases above C_2 the excess chemical present can become toxic and the toxic effect can increase with the dose or concentration. Finally at relatively high doses death occurs. This situation occurs with essential trace metals such as copper, zinc, magnesium and so on.

Many substances are not required for the growth of organisms in their normal life processes. For example DDT, parathion and other pesticides have no known essential life function within organisms. If the concentration of the toxicant is at zero then there is no effect on the organism and no harmful results. However, as the toxicant concentration or dose increases, at some level adverse effects will occur and finally at higher levels death will result.

The exposure period will clearly affect the actual levels at which toxic effects occur for both the essential and nonessential substances. With some ecotoxicants, the concentrations can be relatively high leading to adverse effects over short time periods of exposure. Effects which can be readily observed over short time periods are mortalities. For example, in oil-spill situations there is a very high exposure to the toxicant and large numbers of deaths can occur within 24–48 h. But as a general rule, ecotoxicants occur at relatively low levels and the exposure is for relatively long periods of time. Table 2.4 compares the effects on a population of a dose or concentration over

Table 2.4 Range of effects of a toxicant on a population.

		Dose or concentration	Exposure period	Predicted response
		Very low	Very long (many years)	No detectable effects
		Low	Long (months/years)	Death of sensitive individuals Sublethal effects in survivors
		Intermediate	Intermediate (days)	Equal numbers of deaths and survivors Severe effects in some survivors
		High	Short (hours/days)	Few resistant individuals survive
Increasing dose	Increasing exposure period	Very high	Very short (hours)	Death to all members of the population

a range of levels for various periods of exposure. At one end of this spectrum there is a very high level and short exposure leading to the death of all members of the population. This is comparable to the oil-spill situation. At the other end there is a very low dose or concentration and long periods of exposure which lead to low detectable effects or possibly no effects at all. Exposure to ecotoxicants usually occurs at the low to very low end of the spectrum which means that the responses which are expected in the population would be death of some sensitive individuals and sublethal effects. In ecosystems the same effects would be operating but over a range of organisms together with the added interaction of organisms to give the overall effects. So the effects on ecosystems can be complex and it can be difficult to establish the cause and its relationship to low exposure over long time periods.

Conclusions

Ecotoxicants can arise from many sources in human society. As a general rule ecotoxicants occur as low-level contaminants in larger discharges such as sewerage or stormwater but can occasionally occur in more concentrated forms in situations such as oil spills. It is useful to consider the properties of ecotoxicants in evaluating the effects on ecosystems. A toxicant can be considered to have a range of properties which affect its interaction with ecosystems. Firstly the physical state, solid, liquid or gas, influences the dispersal of the toxicant and its physicochemical properties influence the particular phases which a toxicant will favour within the environment. The biochemical and physiological properties then influence the uptake of a chemical by biota and the interaction with target sites within biota to give lethal and sublethal effects. Ecotoxicants can be classified into groups based on biological effect, chemical properties and physicochemical properties. There is considerable overlap in the usage of these groups and often a combination of terms associated with these groups is used to describe the most important properties of a particular ecotoxicant. The classification into groups allows a better understanding of the properties of individual compounds by consideration of the related properties of the remainder of the group. In addition it allows the properties of new compounds which may fit that group to be indicated by the general properties already known. This can then serve as a basis for further investigation and evaluation of the potential ecological effects of a chemical.

Further reading

Calow, P. (1994) *Handbook of Ecotoxicology*, Vol. 2. Blackwell Science, Oxford.
Connell, D.W. & Miller, G.J. (1984) *Chemistry and Ecotoxicology of Pollution*. John Wiley and Sons, New York.
Schüürmann, G. & Markert, B. (1998) *Ecotoxicology, Ecological Fundamentals, Chemical Exposure, and Biological Effects*. John Wiley and Sons, New York.

3: Distribution and Transformation of Chemicals in the Environment

Introduction

In ecotoxicology we must know the exposure of biota in the system to the chemical to be able to interpret the response. This basic principle applies no matter what chemicals we are considering or their resultant and effects, and irrespective of the biota involved. In fact, in many investigations and research in ecotoxicology a lack of the knowledge of the exposure to the toxic agent involved has made the interpretations effectively impossible. Thus a knowledge of exposure is fundamental to understanding the ecotoxicology of an agent in natural ecosystems. To be able to do this we must be able to evaluate the distribution processes of a chemical in the environment. Firstly, it should be recognized that chemicals are not static in a particular sector of the environment but move physically in the environment under the influence of the movement patterns of air, water, soil and so on and other physical processes such as diffusion. While these processes are occurring chemical transformations may be occurring simultaneously. However, after a period of time, these physical and chemical processes may reach an equilibrium or steady-state situation. The transformation processes in the environment can degrade a chemical to simpler products with less biological impact or, in some less frequent cases, transformation into a product with similar or sometimes greater ecological impact may occur.

After distribution in the environment and ongoing transformation processes have occurred, biota in the system will be exposed to the distributed and transformed chemical. For example, a chemical contaminant discharged into water will be distributed in this environment between various segments or phases. Suspended sediment, the atmosphere and biota within these phases, will be exposed to the distributed chemical. Thus filter feeders will be exposed to chemical on suspended matter as well as that dissolved in the water mass. Also biota in the sediment will be exposed to chemical which has been distributed from the water into the sedimentary mass. However, for a chemical to have a biological impact it must enter biota. The specific entry to biota in the system is influenced by the bio-availability of the chemical and pathways of exposure of the biota to the chemical. For example, in the terrestrial environment a substance entering soil may be very strongly sorbed to the soil and thus not readily dissolved to allow entry to an organism. The entry pathways into organisms through water, soil or sediments, food and so on also influence the bio-availability of the contaminant. After entry into the biota through the various pathways the chemical may arrive at the target site and initiate a reaction

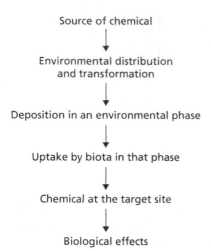

Source of chemical

Environmental distribution
and transformation

Deposition in an environmental phase

Uptake by biota in that phase

Chemical at the target site

Biological effects

Fig. 3.1 The pathways from source of a chemical to biological effects.

of the individual organisms involved. This may subsequently be transferred into higher levels of organization in populations, communities and whole ecosystems. These processes are described in Fig. 3.1.

Environmental partition processes

In considering the environmental distribution of a chemical we must first evaluate the environment itself and simplify it so that we can then consider the environmental distribution process. The best way to simplify the environment is to divide it into phases. A phase is a distinct part of the environment which can be considered to be homogeneous and in which a chemical can be considered to behave in a uniform manner. The atmosphere fits this definition very well as does water in the oceans, lakes, rivers and so on. But we can also consider soil, aquatic sediments, vegetation and aquatic biota to be phases as well. Although these phases are often heterogeneous, a chemical can nevertheless be considered to behave in a uniform manner, and be distributed uniformly, within them.

When a chemical enters the environment it partitions between all these phases to a lesser or greater extent. More of it may go into certain phases than others and, in fact, some phases may take up very little. In simple terms this means the chemical separates into parts which move into the different phases and the amount involved is related to the nature of the phase and the nature of the compound. For example, *fat loving* or *lipophilic* chemicals, because their preference is to dissolve in fat, will move into phases high in fat or lipid and will be reluctant to enter phases with little or no fat. Thus DDT, a lipophilic compound, would be expected to move into lipid-rich phases, such as fish, but only relatively low amounts would enter lipid-deficient phases such as water and air.

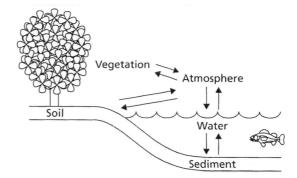

Fig. 3.2 Phases within an environment and the two-phase exchange processes expected to operate.

When a chemical enters the environment there is active exchange between the phases, particularly from the phase or phases in which the chemical is initially discharged. Of course a phase can only exchange chemicals with a phase in which it is in direct contact. So exchange would occur between air and water, water and sediment, and so on as shown in Fig. 3.2. On the other hand, there would be no direct exchange between phases that are not in contact such as fish and vegetation, atmosphere and sediment, and so on. But a chemical can move in a stepwise fashion from phase to phase so that all phases are effectively interacting. We can consider these exchange processes to occur by diffusive movement of the chemical backwards and forwards as shown by the arrows indicating these movements in Fig. 3.2. Thus all of the exchange processes between phases in environmental systems can be represented by sets of two-phase processes. So the processes which occur in the environment shown in Fig. 3.2 can be represented by the following two-phase exchange processes:

vegetation/atmosphere
soil/atmosphere
atmosphere/water
sediment/water
biota/water.

These are the major processes occurring but less important processes could be included in an environment, as represented in Fig. 3.2. For example, the water environment could be expanded to include the suspended sediment/water process. This could make our diagram represent the actual environment more accurately, depending on the levels of suspended sediment present, and thus give a better representation of the system. So the set of two-phase processes indicated should be representative of the actual environment in which we are interested.

We can understand the distribution of a chemical in an environmental system by developing an understanding of how the two-phase systems operate. For these purposes the two-phase systems can be best understood by applying the Freundlich equation:

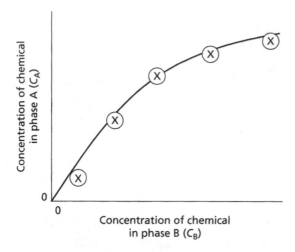

Fig. 3.3 Plot of the equilibrium concentrations of a chemical in two phases (A and B) at different concentrations.

$$C_A = K.C_B^{1/n} \qquad\qquad (3.1)$$

where C_A and C_B are the concentrations in phase A and B, respectively, K is the partition coefficient and n is a nonlinearity constant. The values for this equation can be obtained by taking samples of the two phases, placing the two samples in a set of inert containers, such as glass and inserting varying amounts of the test compound. After time has elapsed and equilibrium occurs then the two phases can be subjected to chemical analysis yielding a set of pairs of concentrations in phase A and the corresponding concentration in phase B. We can evaluate when equilibrium has occurred because at this stage the concentrations in phase A and phase B are constant. The data on the pairs of concentrations can be plotted on a graph as the concentration in A versus the concentration in B as shown in Fig. 3.3. This plot may yield a curved or a straight line and should go through the origin or close to it. When the line is straight then the nonlinearity constant, n, is unity and when the line is curved the value of n deviates from this value. In environmental applications the concentrations of the compounds are usually relatively low. When dealing with concentrations close to zero the line is linear, or we can assume it is (see Fig. 3.3). In this case it can be assumed that n is 1 and thus the reciprocal of n is also 1 and

$$K = C_A/C_B.$$

The application of this equation is dependent on equilibrium being attained which is indicated by the values for C_A and C_B being constant in any particular experiment or situation. It also depends on the same chemical entity moving backwards and forwards to establish the equilibrium. Thus chemical reactions of the chemical under investigation should not occur in either phase. The processes of deposition of a chemical in a phase should be as a result of sorption processes. This means that the processes should be reversible uptake on solid

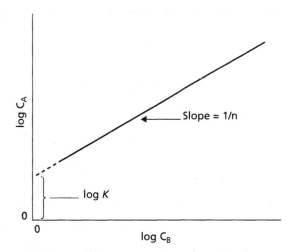

Fig. 3.4 Plot of the logarithmic form of the Freundlich equation.

surfaces, usually referred to as adsorption, or absorption into the phase which often occurs with nonsolid phases.

If the concentrations are not very low, as has been assumed in the treatment immediately above, then the Freundlich equation can be converted into a more readily usable form by taking logarithms of both sides of Equation 3.1. Thus

$$\log C_A = 1/n \log C_B + \log K.$$

This means that plots of $\log C_A$ against $\log C_B$ can be made, as in Fig. 3.4, to readily evaluate $1/n$ and K. In this plot $\log K$ is the intercept of the y-axis when $\log C_B$ is 0 and $1/n$ is the slope of the line which should be linear for all relationships.

Aquatic partitioning processes

The major aquatic partitioning processes are shown in Fig. 3.2. Other processes can be included, if these are of interest, such as those involving suspended sediments and bottom sediments. The two major processes occurring in aquatic systems are:
1 aquatic biota/water;
2 bottom sediments/water.
Both of these processes are reasonably well understood with persistent neutral organic chemicals such as the chlorohydrocarbons.

The aquatic biota/water system

The movement of persistent neutral chemicals dissolved in water into aquatic biota involves passage through the gills during respiration then entry into the

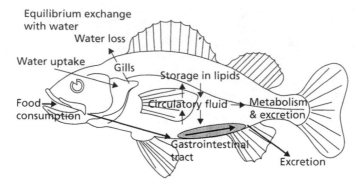

Fig. 3.5 Uptake, accumulation and loss processes for a toxicant in the ambient water with fish.

circulatory fluid and finally deposition in body fat. Since we are only considering persistent neutral organic chemicals, then metabolism and excretion can
be ignored. Also the reverse process occurs as well, releasing chemicals to the
water mass. Uptake from food is generally considered to be a relatively minor
process. These processes are shown in Fig. 3.5. In this way an equilibrium can
be established essentially between biota lipid and the ambient water. If we
assume that equilibrium is with the biota lipid then

$$K_{BL} = C_{BL}/C_W$$

where K_{BL} is the biota lipid/water partition coefficient, C_{BL} is the concentration
in biota lipid and C_W is the concentration in water. But

$$C_{BL} = C_B/y$$

where C_B is the concentration in the whole biota and y is the lipid fraction.
Since

$$K_{BL} = C_B/y.C_W$$

if octanol is considered to be equivalent to the biota lipid and remembering
that C_B/C_W is in fact K_B, the bioconcentration factor, then

$$K_B = y.K_{BL} = y.K_{OW}.$$

The K_{OW} values for a range of different types of compound are shown in
Table 3.1 and applying this relationship to fish with a lipid content of 5%
($y = 0.05$) yields the calculated K_B values shown in this table. It can be seen
that there is reasonable agreement between the measured and calculated
values. We would not expect the agreement to be perfect as variations can be
caused by different species of fish, different sizes of fish and other biological
factors.

Table 3.1 Some values of properties of environmental importance for various compounds.

Compound	Henry's law constant*		K_{OW}	Measured K_B (fish)	Calculated K_B	Measured K_{OC}	Calculated K_{OC}
	Dimensional (H_D) atm-m^3 mol^{-1}	Dimensionless (H)					
Lindane	4.8×10^{-7}	2.2×10^{-5}	5250	470	263	1080	2153
DDT	3.8×10^{-5}	1.7×10^{-3}	2 290 000	1 100 000	1 145 000	243 000	940 000
Arochlor 1242	5.6×10^{-4}	2.4×10^{-2}	199 600	3200	5480	5600	44 700
Naphthalene	1.15×10^{-3}	4.9×10^{-2}	3900	430	199	1300	1600
Benzene	5.5×10^{-3}	2.4×10^{-1}	135	13	7	83	55
Mercury	1.1×10^{-2}	4.8×10^{-1}	–	–	–	–	–
Vinyl chloride	2.4	99	–	–	–	–	–

* Lyman et al. 1990.

The sediment/water system

A similar treatment can be used with the bottom sediments/water process. Thus the sediment/water partition coefficient (K_D) can be developed as follows:

$$K_D = C_S/C_W = f_{oc}.C_{SOC}/C_W = f_{oc}.K_{OC}$$

where C_S/C_{SOC} and C_W are the concentrations in sediment, sediment organic carbon and water, respectively, f_{OC} is the fraction of organic carbon present in the sediment and K_{OC} is the sediment/water partition coefficient in terms of organic carbon (C_{SOC}/C_W). It is known that the organic carbon in sediments has a strong capacity to sorb lipophilic compounds but does not have as strong a capacity as fat or lipid. To account for this a fraction of 0.41 is often used and the organic carbon is assumed to be equivalent to octanol with this fraction applied. Thus

$$K_{OC} = 0.41.K_{OW}.$$

Some calculated and measured K_{OC} values are shown in Table 3.1. Agreement is quite good in some cases and in others shows a considerable variation between these values. This variation could be due to a variety of factors including difficulties in obtaining accurate experimental results and factors such as the composition of the sediments under consideration. Utilizing this expression for K_{OC} the expression below can be obtained for K_D:

$$K_D = f_{OC}.0.41.K_{OW}.$$

Using this relationship K_D value for sediments with specific organic carbon contents can be calculated.

Partition processes involving the atmosphere

Chemicals which are ionized cannot evaporate into the atmosphere under normal environmental conditions. But nonionized chemicals, often referred to as neutral chemicals, can form vapours and evaporate into the atmosphere. These neutral chemicals include a wide range of important environmental chemicals, including environmental contaminants such as DDT and other chlorohydrocarbons, petroleum hydrocarbons and dioxins. The atmosphere can participate in partition processes with such phases as soil, vegetation and water.

Of these processes involving the atmosphere, the atmosphere/water process has been studied for a considerable period of time. This process is described by Henry's law which can be stated as follows:

$$p = H.C_W$$

where p is the partial pressure (pascals, Pa), C_W is the concentration in water (mol m^{-3}) and H is the Henry's law constant (Pa.m^3 mol^{-1}). Thus the partition coefficient is the Henry's law constant which can be expressed in dimensional units which may be different depending on the units in the values used to develop the constant. Henry's law is in fact the ratio of the partial pressure of the compound in the atmosphere to the concentration in the water. This rather unusual form of partition coefficient reflects the long history of this relationship which dates from a time period when the concentrations of compounds in water and air were difficult to measure. Henry's law can be expressed in the dimensionless form, H_D, as

$$H_D = C_A/C_W.$$

The Henry's law constant value, H, can be converted to this dimensionless form, H_D, as follows. The universal gas equation can be expressed as

$$pV = nRT$$

where p is the partial pressure, n is the number of moles, R is the universal gas constant, T is the temperature on the Kelvin scale and V is the volume. Thus

$$p = n/V.RT$$

but n/V is the concentration of the compound in the atmosphere, C_A; thus

$$p = C_A.RT.$$

Divide both sides of this equation by C_W, then

$$p/C_W = (C_A/C_W)RT$$

and so

$$H = H_D.RT.$$

Some values for the Henry's law constant are shown in Table 3.1. These values are all relatively low, indicating that the water concentrations are much higher than those in the atmosphere. Lindane exhibits the lowest value indicating that its volatility is low relative to its solubility in water, and DDT and Arochlor 1242 are somewhat similar to lindane. Naptholene and benzene have relatively higher pressure in comparison to their water concentrations. On the other hand, the situation is reversed for vinyl chloride where the concentration in the atmosphere is higher than that in the water. Mercury has been noted here in this table since in this respect it behaves more like a neutral organic compound than a metal.

The atmosphere/vegetation partition coefficient can be simply developed for neutral lipophilic compounds as shown below:

$$K_{VA} = C_B/C_A = (C_B/C_W)(C_W/C_A)$$

where K_{VA} is the vegetation/air partition coefficient, C_B is the concentration in the vegetation and C_A is the concentration in the air. But K_B is the biota/water bioconcentration factor (C_B/C_W) and the concentration in air/concentration in water ratio (C_A/C_W) is the dimensionless Henry's law constant, H_D. Thus:

$K_{VA} = K_B.H_D$ (at equilibrium).

The bioconcentration factor (K_B) can be described by the following expression, as was shown previously:

$K_B = y.K_{OW}$

where y is the lipid fraction in the vegetation and K_{OW} the octanol/water partition coefficient. Thus

$K_{VA} = y.K_{OW}.H_D.$

So this relatively simple equation can be used to calculate the vegetation/atmosphere partition coefficient at equilibrium. The establishment of equilibrium in atmosphere/vegetation systems may be difficult due to the fluctuating nature of the air and atmospheric concentrations. Using this relationship DDT would have a K_{VA} value of 77.9. Few of these values have been measured so it is difficult to compare the observed and calculated values as was done when discussing previous partition processes.

The soil/atmosphere partition process is probably the most complex of these major processes that have been considered so far. Firstly, there is the difficulty of estimating the thickness of the layer of soil which participates in the atmosphere/soil partitioning process. This varies with the physical consistency of the soil and the ability of air to penetrate it and establish equilibrium. Often 10 cm is used for this characteristic. There is also some evidence that the concentration of the water present in the soil, as moisture, can exercise an influence on the partition coefficient. However, an estimate can be made using the following approach.

$K_{SA} = C_S/C_A = (C_S/C_A)(C_A/C_W) = K_D.H_D$ but

$K_D = f_{OC}K_{OC}$ and

$K_{OC} = 0.41.K_{OW}.$

Thus substituting in the equation above:

$K_{SA} = f_{OC}.0.41.K_{OW}.H_D.$

Using this *expression* DDT would have a K_{SA} value of 16.0 in soil with 1% content of organic carbon.

Terrestrial partitioning processes

In a terrestrial system the atmosphere/soil and atmosphere/plant systems are of major importance. The major characteristics of these systems have been previously described. Terrestrial biota, such as birds, mammals and so on differ fundamentally from aquatic biota in their interaction with the environment. With aquatic biota exchange of chemical contaminants with water is a major controlling factor but with terrestrial organisms the corresponding factor is exchange with the atmosphere. However, the atmosphere has limited capacity to accept persistent organic compounds compared with water. Thus this is a relatively unimportant route of loss.

An important difference between aquatic and terrestrial biota is in their ability to degrade persistent organic compounds. Terrestrial biota generally have a relatively strong capacity to induce an elevated concentration of degrading enzymes after exposure to persistent organic contaminants. The enzyme system induced by this process is referred to as the mixed function oxidase (MFO) system. This enzyme system is able to catalyse the oxidation of persistent lipophilic contaminants to oxidation products which can be relatively easily excreted by organisms. The major route of uptake of persistent organic compounds for most terrestrial organisms is through food and if by means of this route an elevated concentration of compound occurs in the consumer as compared to that found in the food, this process is described as *biomagnification*. The types of organisms involved are the major groups of terrestrial organisms, such as birds, mammals, insects and so on. Thus chemical residues in food enter the gastrointestinal tract and are exchanged through its walls into the circulatory fluid. The metabolic degradation of the residues carried in the blood can occur in various body organs, yielding oxidized products, principally through the MFO system, which are subsequently excreted. Alternatively, lipophilic residues can be deposited from blood into body fat and reactivated from this site under appropriate conditions. Also a proportion of the residue may remain in the digested food and be excreted unchanged. The ratio of concentration in food (C_F) to the concentration in the body of the consumer (C_B) is described as the biomagnification factor (BF). Some measured BF values are shown in Table 3.2. It is difficult to predict the level of biomagnification of a chemical residue from food in the body of a consumer, mainly because of the difficulties in predicting the level of biodegradation of the residues which can occur. Of course biodegradation will reduce the biomagnification factor in proportion to the extent to which it occurs.

Table 3.2 Some biomagnification factors for terrestrial organisms.

Biota	Compound	BF (concentration in body/concentration in food)
Beef cattle	Dieldrin	0.13–3.95
Cattle	Dieldrin	0.11–0.17 (lipid weight)
Cattle	Heptachlor	0.16
Rats	Chlorohydrocarbons	0.02–18.5
Cattle	Chlorohydrocarbons	0.3–2.7

Fugacity modelling of chemicals

The previous discussions are concerned with the behaviour of chemicals in terms of two-phase partition processes. When the environment is considered as a whole there are usually many phases involved. Thus there is a need for a way of considering many phases together to give the ability to integrate sets of two-phase processes into an overall scheme. The approach developed to carry out this operation is based on the *fugacity* of a chemical in a phase.

Fugacity is not a new concept but Mackay (1979) first applied it to modelling the behaviour of chemicals in the environment in 1979. Fugacity is related to pressure and can be conceived as the *escaping tendency* of a substance from any given phase. For a single substance it will vary from phase to phase depending on the characteristics of the substance and the characteristics of the phase itself. Chemicals tend to move from phases in which they have a high fugacity, in other words a high pressure, to those phases where the fugacity of the chemical is low, i.e. having a low pressure. When equilibrium is attained the fugacities in each phase are equal, as are the *escaping tendencies*. Fugacity has a further distinct advantage in that it is proportional to concentration at the relatively low levels of contaminants which are encountered in the environment. This relationship is described by

$$C = f.Z \tag{3.2}$$

where C is the concentration in the phase, Z is the fugacity capacity constant with units of mol m^{-3} Pa^{-1} and f is the fugacity. A single chemical in each phase has its own particular Z value and in terms of a partitioning model, at equilibrium, chemicals will tend to accumulate in those phases with the highest values.

The fugacity concept can be applied to allow calculations to be made of the environmental distribution of a chemical based on the properties of the chemical and the properties of the phases. If all phases are in equilibrium for a distributed chemical then the fugacities of the chemical in each phase are equal and are given by

$$f = C_i/Z_i = M_i/V_iZ_i = \Sigma M_i/\Sigma(V_iZ_i) = M_{total}/\Sigma(V_iZ_i) \tag{3.3}$$

where M_i is the number of moles of chemical in a phase of volume V_i, C_i is the concentration of the chemical and Z_i is the fugacity capacity constant of the chemical in that phase. Using this equation the prevailing equilibrium fugacity can be determined and from this the amount and concentration of the substance in each individual phase can be calculated. However, to calculate the overall fugacity, f, the total mass involved (M_{total}) expressed as moles must be known, the volume of each phase (V_i) measured and the fugacity capacity constant (Z_i) of the substance in each phase estimated. The fugacity capacity constants can be estimated from basic physical chemical data. The fugacity capacity constant is a constant for a given substance in a given phase and is a physical chemical characteristic of that substance in that phase. The fugacity capacity constants can be calculated using the following relationships:

$$Z_{air} = 1/RT \tag{3.4}$$

$$Z_{water} = 1/H \tag{3.5}$$

$$Z_{sediments} = K_D/H \tag{3.6}$$

$$Z_{biota} = K_B H \tag{3.7}$$

$$Z_{soil} = K_{SA} H \tag{3.8}$$

where R is the universal gas constant, H is Henry's law constant, T is the absolute temperature, K_D is the equilibrium constant between sediment and water, K_B is the equilibrium constant between biota and water and K_{SA} is the equilibrium constant between soil and the atmosphere.

This means that the equilibrium partition coefficients such as K_B, H and S_{SA} should be known. In most situations these constants are not known but can often be calculated using quantitative structure activity relationships (QSARs). QSARs have been extensively used to predict the behaviour of chemicals in the environment and are described in more detail in the next section. These applications have been most successful with nonpolar lipophilic compounds which generally exhibit negligible biodegradation. At equilibrium, the fugacities in each phase are equal, thus

$$f_{soil} = f_{air} = f_{sediment} = f_{biota} \tag{3.9}$$

There is only one value for the fugacity, $F_{equilibrium}$, and once this has been calculated this value can be applied to each phase, allowing the distribution of the chemical and also its final concentration at equilibrium to be determined. The overall fugacity at equilibrium can be estimated using Equation 3.3. An example of the calculations of distribution which can be made is shown in Table 3.3.

Fugacity models of the type described above provide a method for predicting environmental concentrations for a discharge of a nondegradable lipophilic chemical into a static environment. This provides a valuable basis for comparison of chemicals and the identification of phases in the environment that are likely to be repositories of discharged chemicals. However, if estimations of actual concentrations likely to occur in the environment are required then additional factors need to be taken into account. For example, allowances are needed for the rate of discharge of a chemical and its rate of degradation in different phases, the rate of movement of water and air phases and the removal of chemical from the model boundaries. All of these factors can be incorporated into the model if the relevant data are available. The accuracy of the results obtained is related to how well the data reflect the actual environment.

Table 3.3 Calculated distribution and concentration of some representative chemicals.

Compound	Pentachlorophenol		Phenanthrene		Tetrachloroethane		DDT		Hexachlorobiphenyl	
	Con* (μg kg^{-1})	% (moles†)	Con (μg kg^{-1})	% (moles)	Con (μg kg^{-1})	% (moles)	Con (μg kg^{-1})	% (moles)	Con (μg kg^{-1})	% (moles)
Air	5.7×10^{-2}	5.8	3.0×10^{-1}	76.2	6.1×10^{-1}	99.9	4.6×10^{-3}	0.4	1.0×10^{-1}	7.5
Water	5.2×10^{-2}	0.6	2.4×10^{-2}	0.4	1.8×10^{-3}	4.0×10^{-2}	4.9×10^{-3}	5.0×10^{-2}	2.0×10^{-4}	2.0×10^{-3}
Soil	33.2	81.0	5.5	20.2	4.6×10^{-3}	2.0×10^{-2}	47.5	86.7	44.5	80.0
Suspended sediments	100	0.2	16.6	4.0×10^{-2}	1.4×10^{-2}	4.0×10^{-5}	142	0.2	134	0.2
Sediment	100	11.9	16.6	3.0	1.4×10^{-2}	3.0×10^{-3}	142	12.7	134	11.7
Aquatic biota	265	2.0×10^{-3}	44.0	4.0×10^{-4}	3.6×10^{-2}	4.0×10^{-7}	376	2.0×10^{-3}	354	2.0×10^{-3}
Vegetation	106	0.6	17.7	0.1	1.5×10^{-2}	1.0×10^{-4}	151	0.6	141	0.6

* Concentration in a single phase.
† Percentage distribution of the compound between all phases.

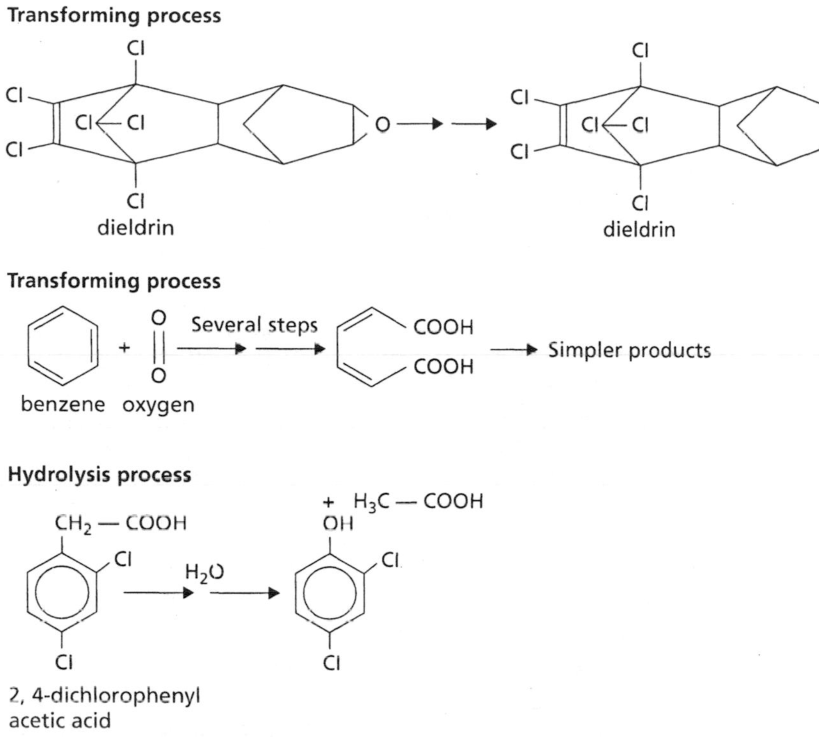
Transforming process
Cl
Cl
Cl
Cl
Cl
Cl
O
dieldrin
O
dieldrin
Transforming process
O
O
Several steps
COOH
COOH
Simpler products
benzene
oxygen
Hydrolysis process
CH2 — COOH
Cl
Cl
H2O
+ H3C — COOH
OH
Cl
Cl
2, 4-dichlorophenyl
acetic acid

Fig. 3.6 Some transformation and degradation patterns for chemicals discharged to the environment.

Chemical transformation and degradation

Chemicals in the environment can be transformed into other closely related forms, with somewhat similar structures and properties, as shown in Fig. 3.6. Chemicals can also be degraded to multiple products of lower molecular weight and this also occurs with transformation products as well. These processes can occur abiotically or be facilitated by biota such as microorganisms.

The major chemical agents in the environment reacting with discharged chemicals are oxygen and water which participate in oxidation and hydrolysis reactions, respectively. Oxygen is a common reactive substance in the environment since it accounts for about 20% of the composition of the atmosphere, and water is also a common environmental component since it is present in the oceans, lakes, rivers and so on as well as within biota. Both oxidation and hydrolysis result in the formation of products containing hydroxyl, carbonyl and other polar groups which result in an increase in polarity of the products and a resultant increase in water solubility. Thus with lipophilic compounds the products have higher polarity and the greater solubility in water results in easier excretion and removal from the organism. Examples of oxidation and hydrolysis reactions with environmental contaminants are shown in Fig. 3.6.

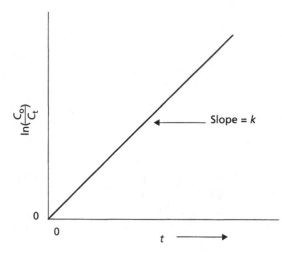

Fig. 3.7 Plot of loss of chemical from an environmental compartment over time according to first-order kinetics where C_0 is the concentration at time zero and C_t the concentration after an elapsed period of t, and k is the rate constant.

The oxidation of benzene and hydrolysis of 2,4-dichlorophenoxyacetic acid yield multiple degradation products containing oxygen which are relatively water soluble and easily excreted by organisms.

The kinetics of transformation and degradation are very important in considering environmental processes since the rate of change of occurrence of environmental contaminants has a powerful influence on the spread of a contaminant and its persistence over time. Generally environmental processes can be considered to follow first-order kinetics, thus:

$$-dC/dt \propto C$$

where C is concentration in the environment and t is the elapsed time. This means that the rate of reaction is proportional to the concentration of the contaminant present in the environmental phases

$$-dC/dt = kC$$

where k is the first-order rate constant with units of time^{-1}. On integration and rearrangement the following equation can be derived:

$$\ln(C_0/C_t) = kt \text{ or } C_t = C_0\, e^{-kt}$$

where C_0 is the concentration of the organic substance at time 0, and C_t is the concentration of the substance after a time period of t. This can be plotted as shown in Fig. 3.7 with $\ln(C_0/C_t)$ on the vertical axis and the time period t, on the horizontal axis. The slope of this plot represents the rate constant k. The values of the rate constant k are in reciprocals of time and thus inconvenient units in which to readily comprehend the meaning of this characteristic. A more convenient measure of the persistence of a chemical in the environment is the half-life $t_{1/2}$. An expression for this can be derived as follows. According to first-order kinetics it can be shown that

Table 3.4 Half-lives of some pesticides in soil.

Compound	Half-life (years)
Chlorohydrocarbons	
DDT	3–10
Dieldrin	1–7
Toxaphene	10
Organophosphate compounds	
Pyfonate	0.2
Chlorfenos	0.2
Carbophenothion	0.5
Carbomates	
Carbofuran	0.05–1

$t = \ln(C_0/C_t)/k.$

Starting at any point in time when half of the original substance has been degraded, then

$C_0/C_t = 1/0.5 = 2.$

Thus,

$t_{1/2} = \ln 2/k = 0.693/k = a \text{ constant}.$

This means that the half-life $(t_{1/2})$ is constant for a given compound and a given environmental degradation process that occurs under specific conditions. This is found to be generally true for environmental processes but there is a considerable level of variability due to natural variations in pH, temperature, availability of water, availability of oxygen and other factors that influence persistence in a particular situation. Some examples of half-life that have been measured are shown in Table 3.4.

Bioavailability

The previous discussion is concerned with the distribution of toxicants in the environment since this leads to the exposure of organisms present in different environmental phases. When exposure occurs often only a proportion of the chemicals in the phase is available to be taken up by organisms. This is usually referred to as the bioavailability of the toxicant. For example, with lipophilic compounds in sediments the bioavailability is strongly influenced by the organic carbon content of the sediment. This effect can be interpreted by the partitioning approach previously described where the amount of chemical available to be taken up is that available in the water phase. With increasing organic carbon content of the sediments more compound is partitioned into the sediments and less is available in the water phase. For many metals, including cadmium, nickel and lead, the acid volatile sulphide (AVS) content of the sediment is a strong factor influencing bioavailability due to the strong reaction capacity of sulphur with metals. AVS is a measure of the extractable fraction of the total sulphide content present associated with sediment mineral surfaces.

In order to evaluate the actual exposure of an organism in a phase the bioavailability needs to be taken into account. This can allow estimates to be made of the entry of the contaminant into an organism which is related to the amount of a chemical present at the actual target site in an organism. This might give a more accurate expression for exposure than the concentration in a phase.

Quantitative structure activity relationships (QSARs)

In ecotoxicology research and evaluation there is often a need for many environmental parameters. The parameters are needed to calculate partition coefficients, the distribution of a chemical in the environment and a variety of other environmental characteristics. However, these properties are often not available and techniques need to be used to make estimates of probable values. On the other hand some characteristics which can be measured fairly readily in the laboratory, such as K_{OW}, aqueous solubility and so on, are more readily available. To cope with this situation relationships have been established between physical chemical properties and environmental properties based on known characteristics, and then these are used to predict properties of compounds which are not available. These relationships are usually referred to as *quantitative structure activity relationships* (QSARs).

QSARs are extensively used in the investigations of the ecotoxicology of chemicals in the environment and have found particular use in predictions of sediments/water distribution, the bioconcentration factor and toxicity. These characteristics are key factors in understanding the behaviour and likely effects of chemicals in the environment. These applications have been most successful with nonpolar, lipophilic compounds, which exhibit negligible biodegradation. In addition the octanol/water partition coefficient (K_{OW}) has been found to be a parameter of major importance as illustrated by the plot in Fig. 3.8. Previously expressions for the sediment/water partition coefficient (K_D), the bioconcentration factor (K_B) and other environmental partition coefficients were developed. These environmental characteristics could be described by QSARs. A set of typical QSARs for various environmental processes is shown in Table 3.5.

Although K_{OW} values have been found most useful in QSARs for bioconcentration, Table 3.5 also includes an equation based on aqueous solubility. Of particular interest are the molecular descriptors, such as molecular surface area, molecular volume and related descriptors as well as the range of connectivity indices, since these can be calculated from the chemical structure without the need for any experimental investigation. An example of one of these, the

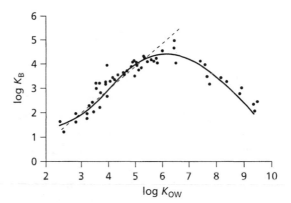

Fig. 3.8 Plot of the approximately parabolic relationship between $\log K_B$ and $\log K_{OW}$ with the approximately linear relationship that exists between $\log K_{OW}$ of 2 and 6.

Table 3.5 Summary of various quantitative structure activity relationships (QSARs) for lipophilic compounds.

Biota groups	Process	QSAR	Factor
Fish	Bioconcentration	$\log K_B = \log K_{OW} - 1.32$	Octanol/water partition coefficient (K_{OW})
Fish	Bioconcentration	$\log K_B = 1.93 - \log S$	Water solubility in moles m^{-3} (S)
Fish	Bioconcentration	$\log K_B = -0.171\,(^2X^v)^2 + 2.253$ $(^2X^v)^2 - 2.392$	The second-order valence molecular connectivity index ($^2X^v$)
Molluscs	Bioconcentration	$\log K_B = 0.844 \log K_{OW} - 1.23$	Octanol/water partition coefficient (K_{OW})
Microorganisms	Bioconcentration	$\log K_B = 0.977 \log K_{OW} - 0.361$	As above
–	Sediment/water	$\log K_{OC} = 1.029 \log K_{OW} - 0.18$	As above
Fish	Toxicity	$\log (1/LC_{50}) = 0.907 \log K_{OW} - 4.94$	As above

second-order valence molecular connectivity index, incorporated into a QSAR of bioconcentration, is included in Table 3.5. These QSARs are generally applicable to chlorohydrocarbons and polyaromatic hydrocarbons, principally as a result of the environmental stability of these compounds. The relationships apply only when equilibrium has been established; the time periods required to establish equilibrium can be unexpectedly extended and in many situations a final equilibrium concentration may not be reached.

Global contaminants

Recent research has shown that certain toxic, persistent and bioaccumulative chemicals, typified by DDT, dieldrin and the polychlorinated biphenyls (PCBs) occur widely throughout the global environment. These substances originate from a variety of sources. Many are agricultural pesticides, although some are industrial chemicals, such as the PCBs, while others are principally combustion products, such as dioxins and furans. There have been considerable changes in the usage patterns of many of these substances in recent years. The production and use of many chlorohydrocarbon pesticides has been restricted or banned in most industrial countries. However, the usage of these and other related compounds is increasing in many tropical countries. It appears that the restrictions and banning of some of the agricultural chemicals has caused a decline in environmental contamination within and adjacent to those countries. On the other hand, there has been an increase in contamination by these substances in many tropical regions. As a general rule the levels of these contaminants in and adjacent to tropical countries tend to be higher than those associated with industrial countries. However, the levels of some of these substances are more uniformly distributed around the globe. For example, the hexachlorocyclohexanes, hexachlorobenzenes and chlordane are widely distributed throughout the global environment. In addition some of these substances exhibit an increase in concentration with locations closer to the polar regions. It is suggested that this is due to the volatilization of these compounds in tropical regions and transfer by atmospheric movements and ocean currents to colder regions where they condense and produce relatively high levels in the natural environment.

Currently, an international consensus is being sought to identify actions that need to be taken with respect to these global contaminants. Increasing evidence is coming forward that these trace substances are having deleterious effects on biological systems; evidence in particular regarding the implication that some of these substances are active in the disruption of endocrine systems leading to a lack of reproductive success. Through long-range transport and deposition on a global scale these substances can contaminate locations far from their source. For this reason actions are being sought for their control as a result of international agreement.

Conclusions

Currently, our understanding of environmental distribution processes has reached a reasonably advanced level and methods for prediction are at a fairly high level. However, it can be expected there will be continued interest in the development of methods for the prediction of distribution of chemicals in the environment. There are a large number of factors which are not understood and this understanding would assist greatly in improving our ability to model these processes. The use of molecular descriptors and other parameters that can be obtained by calculation would be expected to increase since these do not involve expensive experimental work. In addition, many of these investigations could contribute to our basic understanding of the underlying principles, which requires expansion so that prediction methods can be applied more effectively. Nevertheless it should be kept in mind that most prediction methods have been developed based on laboratory experiments and the suitability of these relationships for understanding field situations needs clarification.

Further reading

Connell, D.W. (1988) Quantitative structure activity and relationships and the ecotoxicology of chemicals in aquatic systems. *Atlas of Science* **1**, 221–225.

Connell, D.W. (1990) *Bioaccumulation of Xenobiotic Compounds*. CRC Press, Boca Raton, Florida.

Connell, D.W. (1997) *Basic Concepts of Environmental Chemistry*. Lewis Publishers, Boca Raton, Florida.

Lyman, W.J., Reehl, W.F. & Rosenblat, D.H. (1990) *Handbook of Chemical Property Estimation Methods—Environmental Behaviour of Organic Compounds*. American Chemical Society, Washington, D.C.

Mackay, D. (1979) Finding fugacity feasible. *Environmental Science and Technology* **13**, 1218–23.

Mackay, D. & Patterson, S. (1981) Calculating fugacity. *Environmental Science and Technology* **15**, 106.

Mackay, D. & Patterson, S. (1982) Fugacity revisited. *Environmental Science and Technology* **16**, 654A.

4: Molecular, Biomolecular, Physiological and Behavioural Responses of Organisms

Introduction

Toxicants can enter biological systems via different routes. These substances may be taken up through respiratory surfaces (e.g. gill, lung, trachea), general body surfaces (e.g. skin, cell wall) or oral ingestion (e.g. food, water). It should be noted that, for a given dose, the toxic effects of a toxicant may vary, depending on the route of entry. Respiratory surfaces often have a very large surface area to facilitate gaseous exchange; for example, the total area of alveoli in a human being is some 25 times the total area of the body surface. Because of these characteristics of the respiratory surface, toxicants are much more readily dissolved and absorbed across the cell membrane, and the total quantity of toxicant taken in is significantly higher. The high absorption rate and the high efficiency of transport within the body means a very high dose will be taken up and distributed within a relatively short time. Consequently, the intake of toxicant through the respiratory system generally results in a higher toxicity, and explains why the inhalation of a given dose of a toxicant may be much more toxic than the corresponding percutaneous or oral intake.

Terrestrial and aquatic animals often have a external body surface which is impervious to most toxicants. Absorption of toxicant through the body surface is therefore normally much slower. On the other hand, toxicants entering the body by means of oral ingestion must pass through the alimentary canal before absorption can take place. In this case, the chemical nature, bioavailability and hence toxicity of the toxicant may be changed by digestive enzymes or the modified pH environment in the alimentary canal. In some cases, the chemical nature and toxicity may be altered by microbial activities in the digestive tract. Cycasin is a glycoside found in certain plants which serve as human food sources in the Pacific islands. While cycasin itself is nontoxic, it can be converted into a mutagen, aglycon methylazoxymethanol, by B glycosidase produced by gut microflora.

After entering the biological system, toxicants may be eliminated in several ways. Elimination may be achieved through excretion, transformation into other less toxic forms through biological activities and metabolism, or sequestration in metabolically nonactive tissues. As long as the elimination rate is higher than uptake rate no net accumulation of toxicant will occur within the biological system. Toxicity also depends on the biological half-life of the xenobiotics in the biological system, and toxicity is likely to be low for those chemicals with a short biological half-life.

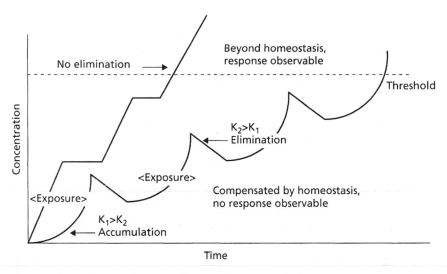

Fig. 4.1 Effects of repeated periods of exposure on toxicant behaviour in an organism.

Provided that the level of toxicant is kept below a threshold within the biological system, referred to as homeostasis, no biological effects may be observable, although energy is required for clearance (see Fig. 4.1). However, if the uptake rate is greater than the clearance rate, toxicants in the biological system will eventually build up in concentration with time. This often occurs during chronic exposure to a chemical. In such cases, the level of toxicant will exceed the threshold value sooner or later. Remember, any chemical will become toxic provided that the dose is high enough!

Toxicants may induce biological responses at different levels of biological organization. Biological responses usually begin at lower organizational levels and gradually manifest themselves at higher organizational levels. At the molecular level, toxicants may bind to DNA, alter its structure and initiate a cascade of effects including cancer development. They may induce or suppress the expression of certain genes and thereby initiate or suppress synthesis of certain protein products which will alter normal molecular functions. At the biochemical level, toxicants may directly induce or suppress enzyme activities, alter essential biochemical pathways and impair normal metabolism by competing with metabolites for active binding sites. Animals and plants may simply tolerate minor alternations, or alternatively, they may restore their normal functions through a variety of homeostatic mechanisms. If molecular and biochemical responses cannot be compensated, molecular and biochemical changes may be progressively expressed at higher levels and physiological changes (e.g. changes in heartbeat, hormone levels, respiration, photosynthetic rate, osmoregulation, etc.) may result. Major physiological disturbance may lead to significant behavioural changes (e.g. in feeding, locomotion, response to light and other environmental factors, etc.). Major physiological or behavioural impairment may affect more obvious traits like growth and reproduction, which may directly or indirectly affect the survival of the organisms or the species in their natural environment. In the extreme event of physiological failure, death will occur. Beyond limits of compensation, sublethal responses at lower organization levels, may progressively express at higher organization levels (population, community and cosystem level).

The normal period required for toxic responses depends on both the nature and dose of toxicant. For sublethal dosages, molecular and biochemical responses may occur within minutes to hours and physiological responses within hours to days, and it may take months or years for symptoms to be manifested as damage in growth and reproduction and lethality. Evidence of significant effects and damage at the population, and community levels and alternation of ecosystem function may not be apparent for years or decades. From an ecotoxicological point of view, it would be obviously very useful if one could predict or extrapolate the occurrence of response(s) at higher organization levels, for example, at population, community and ecosystem levels, from the observed responses at lower organization levels, i.e. from the physiological, biochemical, behavioural, molecular and response of individuals. This would provide a useful early warning system long before environmental degradation could actually occur at the higher levels. In this chapter, we will consider the uptake of toxicants from the environment into the biological system, and the subsequent possible suborganic level responses.

The uptake, metabolism, transformation and elimination of a toxicant depends very much on the nature and structure of the compound. Before a toxicant can reach the receptor site to initiate toxic actions, it must be taken up from the environment into the biological system and transported to the receptor site. Chemical transformations may possibly take place before the toxicant reaches the receptor site. During these transformation processes, toxicants may be metabolized, and their chemical nature and toxicity modified. Alternatively, toxicants may also be excreted outside the body or sequestrated in metabolically inactive tissues. In such cases, toxicants may be reduced in concentration or may never even reach the receptor site in sufficient levels to exert any observable toxicity. These five processes (uptake, transport, metabolism, excretion and sequestration) are described as *toxicokinetic processes*, which is a term borrowed from pharmakinetic evaluations. Toxicokinetic processes determine how much toxicant reaches and interacts with the receptors. The interactions of toxicants with the receptor site, and the consequent induction of toxic effect are called *toxicodynamic actions*. The degree of most toxicodynamic reactions depends on how many receptor sites are bound to or react with toxicant molecules.

Toxicokinetic processes: uptake, transport, metabolism, sequestration and excretion

Uptake

A toxicant can enter unicellular organisms simply through direct passive diffusion across the cell membrane. Thus with multicellular macroalgae, the toxicant is taken up through simple diffusion. For terrestrial plants, toxicants dissolved in water or soil may be taken up through roots or leaves, while gaseous toxicants and aerosols in air may enter plants directly through stomata on the leaves. Lipophilic compounds such as herbicides may penetrate the wax cuticles on the plant surface. In animals, toxicants may enter the body via three important routes. First, toxicants may be taken up by ingestion of food and drinking water. In such cases, toxicants have to pass through the epithelial cell membrane of the digestive tract, and the nature of the toxicant (and hence its toxicity) may be subject to modification by digestive enzymes, pH and microbial activities associated with the digestive tract. Second, toxicants may be taken up through respiratory surfaces, i.e. tracheae in insects, gills in aquatic animals and the lung in terrestrial animals. Third, some toxicants may be taken up directly through the general body surface.

Regardless of their route of entry, toxicants must first pass through a cell membrane or cell wall separating the internal biological environment from the external environment.

The major barrier to uptake of toxicant is the cell membrane or cell wall at the site of uptake, e.g. cell membrane of skin or gastrointestinal tract, epithelial cell of gill and tracheae, cell wall of plant root, etc. Four principal mechanisms may be involved in movement of toxicants across the cell membrane. These include: passive diffusion, facilitated diffusion, active transport and pinocytosis.

Passive diffusion

Passive diffusion involves random movement of molecules across the cell membrane along a concentration gradient until equilibrium is reached. This uptake process is the simplest form and also the most common for foreign compounds, especially for uncharged molecules. Although the process of passive transport is not substrate specific, different transport processes are involved, depending on the chemical and physical nature of the toxicant molecule. Inorganic and organic toxicants with a small molecular size (< 0.4 nm) and dissolved in water can move across pores on the membrane along a concentration gradient. For example, uptake of carbon monoxide (CO), prussic acid (HCN) and nitrous oxide (N_2O) in the lung is by passive transport. Lipophilic chemicals, on the other hand, may diffuse through the lipid bilayer in the direction of the concentration gradient. Rate of passive transport is a first-order rate process, and depends on the following factors.

1 Concentration gradient at the cell membrane.
2 Surface area and thickness of the membrane.
3 Lipid solubility and ionization of the toxicant to be transported.
4 Molecular size.

Facilitated transport

A number of transmembrane carriers or protein pores exist on the cell wall and may greatly enhance (e.g. up to 50 000-fold) the transport of certain classes of molecules across the cell membrane in the direction of the concentration gradient. This process is energy independent, and called *facilitated transport*. Transport of calcium across a membrane, for example, is facilitated by a calcium-binding protein, calmodulin.

Active transport

Toxicants may be carried across the cell membrane by active transport. This process requires energy expenditure and is independent of, or against, concentration gradient. Transport of toxicants across the cell membrane is mediated through a carrier or a pump system. The toxicant molecule is first combined with the carrier on the outside of the membrane, and then detached from the carrier and deposited inside the cell. Active transport is often substrate specific, and a carrier only combines with certain toxicants. The rate of active transport is governed by kinetics of enzyme reactions, and is limited by the number of carriers on the membrane. Polar molecules which cannot be taken up by passive diffusion may be taken up by active transport. The P-glycoprotein pump, for example, is involved in transporting a number of xenobiotics (e.g. organochlorines) out of the cell in the detoxification process. The calcium pump, in which Ca^{2+}-ATPase is involved, is responsible for transporting calcium ions across the cell membrane.

Pinocytosis

Large-size molecules cannot pass through the cell pores of the membrane, and must be transported across the membrane through *pinocytosis*. Pinocytosis involves infolding of cell membrane around the toxicant molecule outside the cell and release of the molecule inside the cell (endocytosis). Transportation of small particle size (< 1 µm) airborne toxicants across the alveolar cells of the lung is carried out by this process.

Transportation within the organism after uptake

After entering the biological system, toxicants can be transported by blood, lymph or haemolymph and distributed to other parts of the organism. In

Table 4.1 Physiological targets of some common toxicants.

Toxicant	Target areas
Lead	Bone, teeth
Mercury	Nervous tissue
Organochlorine pesticides, polychlorobiphenols (PCBs)	Fat, milk
Aflatoxin, vinyl chloride	Liver
Asbestos	Lung

terrestrial plants, transportation is by the water stream in the xylem, or cytoplasmic strands called plamadesmata in the phloem. These two translocation processes in plants are highly dependent on environmental factors such as temperature, air humidity, light quality and intensity and moisture content of soil, and so consequently is the transport of toxicants. Due to the chemical characteristic of toxicants and tissue types, certain toxicants may have a great affinity for certain types of tissue or organs. Such chemical affinity leads to the deposition and localization of toxicants in certain *target organs*, where they may be stored or exert their specific toxicity. For example, cadmium (Cd) and lead (Pb) taken up by the scallop *Pecten alba* are mainly sequestrated in gonads but not in adductor muscles. In mammals, mercury (Hg) tends to deposit in different organs following the order: kidney $\rightarrow$ liver $\rightarrow$ spleen $\rightarrow$ intestine $\rightarrow$ heart $\rightarrow$ muscle $\rightarrow$ lung. Cadmium tends to concentrate in kidney and bone, while Pb tends to deposit in brain tissue and bone. Due to their lipophilicity, most organic xenobiotics are deposited in fatty tissues. The nervous system is the target organ for organophosphorus pesticides where they affect neurotransmission by inhibiting acetylcholinesterase activity. The physiological targets of some common toxicants are shown in Table 4.1. There are numerous examples of organotrophy, but the mechanisms of organ-specific toxic effects remain obscure.

Transformation

Living organisms have had to deal with a great variety of toxicants since life began. These toxicants may be their own endogenous metabolites or naturally occurring toxicants in the living environment. For example, many organisms —e.g. bacteria, fungi and sponges as well as higher plants and animals— produce a great variety of biotoxins as a defensive mechanism to deter predators and competitors. Aflatoxins produced by moulds, toxins produced by wild mushrooms, tertadoxins produced by puffer fish, endotoxins such as botulinus toxins produced by bacteria are among some of the best-known examples. At later evolutionary stages, additional cytochrome families of detoxifying enzymes may have evolved to metabolize toxic compounds such as polycyclic aromatic hydrocarbons (PAHs), produced as combustion products. PAHs can result from forest fires and existed in the natural environment long before human beings appeared on this planet. At the same time, this unique enzyme system also serves to detoxify many endogenous metabolites produced by the living organisms themselves, such as prostaglandins, steroid hormones, fatty acids and bile salts. More recently, human activities have generated a great variety of new foreign chemicals, and the same enzyme system has further evolved and adapted to cope with these xenobiotics. Plants may deal with toxicants by increasing the activities of a variety of oxidative enzymes—e.g. cytochrome oxidase, phenol oxidase, peroxidase, ascorbic acid oxidase—to transform and oxidize toxicants.

One of the basic strategies of detoxification by organisms is to eliminate the toxicant from the body as quickly as possible in order to minimize exposure. Since most of the main excretory organs have evolved to deal with water-soluble compounds, dissolving the toxicant in water greatly facilitates its elimination through this readily available excretory system. In vertebrates, chemicals often must be dissolved in urine or sweat before they can be excreted and similar processes are involved with other organisms. To convert toxicants into more water-soluble forms for elimination from the organism, the general strategy is to convert the toxicant molecules into more polar, hydrophilic compounds. In most organisms, this is normally achieved through two sequential phases described as Phase I and Phase II transformations. In Phase I transformations, a polar group is normally added to the toxicant molecule to increase its polarity and pave the way for further chemical reactions in Phase II transformation. In Phase II transformations, a variety of endogenous metabolites (e.g. sugars, peptides or sulphate) may be conjugated to the Phase I metabolic product by formation of covalent bonds. This increases the polarity and hydrophilicity of the toxicant molecule and so enhances its elimination.

Phase I transformation

The enzyme system responsible for Phase I transformations is known as the mixed function oxidase (MFO) system, or also known as mixed function oxygenase. MFO enzymes are membrane-bound proteins located in smooth endoplasmic reticulum, and extractable in the form of membrane vesicles called microsomes in the laboratory and is also known as microsomal oxidase. The MFO system is highly conservative, and is known to occur in all animal and plant species studied thus far.

The P450 gene superfamily, which initiates MFO formation is known to contain more than 150 genes, and is widely found in plants, animals and prokaryotes. In vertebrates, these enzymes are primarily found in the liver parenchyma cells (which explains why the liver is the main organ for detoxification in vertebrates), although the enzyme family is also found in intestines, gills and other tissues. In molluscs and crustaceans, these enzymes are present in hepatopancrease and digestive glands. In general, MFO activities are lower in molluscs and this explains why certain toxicants such as PAHs can be accumulated in molluscs but not mammals.

The following generalized reactions indicate some of the processes catalysed by the MFO system:

Hydrolysis
$$RCOO-R' + H_2O \longrightarrow RCOO-H + R'-OH$$

Epoxidation

$$R-CH-CH-R' \longrightarrow R-\overset{\displaystyle O}{\overset{\displaystyle /\backslash}{CH-CH}}-R'$$

Dehalogenation
$$R-Cl \longrightarrow R-H + Cl^+$$

Phase II transformation

The next stage (Phase II) of biotransformation is to conjugate the Phase I metabolites of these xenobiotics to a range of water-soluble endogenous metabolites. Cytochrome P450 II enzymes are responsible for these chemical reactions. Chemical reactions at this Phase II include deamination, acyclic hydroxylation, aromatic hydroxylation and dealkylation. The end result is to increase polarity of these xenobiotics to facilitate excretion in urine, bile and sweat. It should be noted that, however, some of the metabolized intermediate products may be more toxic or carcinogenic.

While the endoplasmic reticulum is the site for Phase I transformation, Phase II transformation reactions take place in cytosol. In Phase II transformation, a variety of endogenous metabolites such as sugar, amino acids or their derivatives (e.g. glucuronic acid, glutathione), phosphates and sulphates may be conjugated to the Phase I metabolites by covalent bonding. These polar endogenous metabolites are normally added to the functional group or groups originally on the toxicant or introduced by the Phase I reaction. The Phase II conjugation will further increase the polarity and water solubility of the toxicant, thereby decreasing the liphophilicity of the compound, facilitating subsequent excretion. Important Phase II enzymes include glutathion S-transferases (GST), epoxide hydrolase (EH), UDP-glucuronosyltransferase (UDP-GTS) and sulphotransferase (ST).

Induction of the MFO system

Specific forms of MFO enzymes are inducible by exposure to a variety of lipophilic xenobiotics, e.g. organochlorines, polychlorinated dibenzodioxins, polychlorinated dibenzofurans, PAHs and PCBs. This explains why exposure of animals to sublethal levels of toxicants may increase tolerance of the organisms to the chemical. As a result, elevation of MFO enzymes may be used as a biomarker to indicate that animals have been exposed to these xenobiotic compounds. In fish and mammals, induction of the CYP1A system involves binding of the xenobiotics to the aryl hydrocarbon (Ah) receptor. The xenobiotic –receptor complex then binds to a translocating protein which enters the nucleus and binds to a specific site of the DNA, resulting in transcription for CYP1A enzymes. Induction may be very rapid (within a few hours), with increased activities of CYP1A of some 100-fold. Likewise, Phase II enzymes in fish are also inducible up to several-fold after treatment with PCBs and PAHs. Since MFO is also responsible for metabolizing a variety of endogenous metabolites, it should be noted that the induction of MFO is not specific to anthropogenic toxicants. For example, it has been reported that MFO is also inducible by certain endogenous factors, e.g. hormones, and environmental factors, e.g. temperature stress.

In plants, xenobiotics may be metabolized by oxidative metabolism. The main enzyme responsible is the peroxidase, cytochrome P450-related enzyme (mainly cinnamic acid 4-hydroxylase). This enzyme system is inducible by herbicides.

Although biotransformation by the MFO system may detoxify many xenobiotics giving protection to the organism, biotransformation does not always mean detoxification. Indeed, MFO biotransformation may also convert some chemicals into much more toxic metabolites which can exert harmful effects. For example the common polyaromatic hydrocarbon, benzo[a]pyrene is not a carcinogen itself, but MFO converts benzo[a]pyrene into benzo[a]pyrene diol epoxide. This metabolite is a strong electrophile which may bind to the guanosine moiety of DNA, causing point mutation and oncogene activation. Thus benzo[a]pyrene is a strong carcinogen and mutagen through biotransformation as shown in Fig. 4.2. Since some of these biotransformations are inducible by specific toxicants, elevation of levels of P450 enzymes (especially 7-ethoxyresorufin O-deethylase—EROD) in the livers of feral and cultured fish

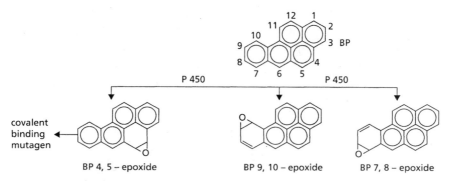

Fig. 4.2 The conversion by mixed function oxidase (MPO) action of the noncarcinogen polyaromatic hydrocarbon, benzo[a]pyrene, into benzo[a]pyrene diol epoxide which is a strong carcinogen.

has been used to monitor trace organics, especially chlorophenols, in water. EROD activity is directly related to levels of pulp-mill contamination, and the effect is observable some 13 km downstream of a pulp mill.

Vertebrates have a much more efficient MFO system in their liver, thus chlorohydrocarbons, PAHs and other lipophilic compounds can be rapidly metabolized into excretable water-soluble products and do not bioaccumulate in human beings. Molluscs, on the other hand, have low MFO activity, and lipophilic contaminants entering the body will tend to bioaccumulate.

Sequestration

Another protective mechanism which animals have evolved to deal with toxicants is sequestration. Storage of toxicants, especially storage in inert tissues such as fat, teeth, hair and horns, means these substances are removed from the general circulation and their toxicity is reduced. In plants, toxicants may be deposited in vacuoles, leaves or bark followed by shedding. Thus, no toxicity will occur until the storage receptors become saturated, or when these toxicants are displaced by some other chemicals which have a higher affinity for the same storage receptor. For example, antidiabetic sulphonylureas may be displaced by sulphonamides. Due to their lipophilic nature, this category of chemicals (chlorohydrocarbon pesticides, organochlorines and PCBs) may be deposited in inert tissues such as fat and hence become inert in the biological system. However, toxic effects may occur when lipid reserves of the animal are catabolized during adverse conditions, e.g. during starvation. At this time, toxicants stored in fat will be released into the bloodstream and reach a high concentration within a relatively short period of time and toxic effects may become evident. Due to their lipophilicity, chlorohydrocarbons in mammals may be 'stored' at high concentration in milk, and breast feeding may pass the contaminants to the young, thereby exerting a deleterious effect on the latter. For example, it has been shown that chlorohydrocarbon pesticides and PCBs in female seals may be passed to their young through this mechanism. Animals may also deposit certain toxicants, e.g. lead, in their inactive tissues such as hair, nail and horns resulting in a reduction of toxic effects. Barnacles, for example, may deposit zinc in their gut epithelial cells in the form of zinc granules, reducing the bioavailability of the toxicant and hence its toxicity. The patterns of toxicant sequestration in animal and plant tissues are often used by ecotoxicologists to monitor specific pollutants in the environment.

Metallothionein (MT) and MT-like proteins are perhaps among the best-studied proteins in the field of toxicant sequestration. Metallothionein is a low molecular weight (6000–10 000 Da) cytoplasmic protein which is very rich in cysteine (up to 30%). MT is widely reported in a great diverse group of eukaryotes. It has a strong binding affinity for a variety of metals. It is widely found in animal and plant taxa and is inducible by exposure to metals. Significant correlation has been found between levels of various metals, including Cd, Zn and Hg, and MT levels in different types of mammal tissues. There is no agreement on the biological significance of this protein but it is probable that MT evolved to regulate the availability of essential metals within the cell. Free metal ions entering the cytoplasm will induce the synthesis of MT and bind with high molecular weight (e.g. metalloenzyme) and low molecular weight (phytochelatin) ligands and MT. The biological half-life of MT in the mussel *Mytilus edulis* is about 25 days. Recent studies have shown that MTs are inducible not only by metals, but also by other stresses such as hypoxia and temperature changes. In plants, phytochelatin may be produced, and the function of phytochelatin is similar to that of MT. The binding of metals to MT or phytochelatin serves to regulate both essential and non-essential metals within the cell and reduce their interaction with other macromolecules.

Excretion

The rate of elimination of xenobiotics from an organism plays an important role in controlling the level of toxicity which occurs. The longer the retention time, measurable by the biological half-life, of the chemical in the organism, the more likely it is to exert a significant toxic effect. Since the major routes of excretion in plants and animals are designed to handle water-soluble metabolites, lipophilic toxicants must be metabolized into hydrophilic compounds before excretion.

In vertebrates, nongaseous and nonvolatile compounds are normally excreted through the kidney. The blood passes through the glomerulus where filtration takes place. The pore size of the glomerulus membrane is fairly large (40 Å), and permits compounds with a molecular weight < 70 000 to pass through. Both active uptake and passive diffusion are involved. Ionized, nonprotein-bound compounds are excreted by active secretion while other compounds may be excreted by passive diffusion. Lipid-soluble, nonionized compounds are reabsorbed into the bloodstream by passive diffusion.

Bile is another important excretion route for xenobiotics, and this process mainly involves active transport. Compounds excreted through bile mainly consist of conjugates with a high molecular weight (generally < 300). The excreted conjugates in the bile may be hydrolysed by gut microflora, and the xenobiotics, or their metabolites, released may then be reabsorbed in the intestine and transferred back to the liver via the enterohepatic circulation. Since active transport is mainly involved in bile excretion, saturation of plasma protein-binding sites of the active transport system may exceed the threshold values and lead to expression of toxicity.

Gaseous (e.g. ammonia) and volatile (e.g. alcohol) compounds are usually eliminated from the gill and lung by simple diffusion. Other minor sites of excretion may include milk, sweat and saliva. In mammals, highly lipid-soluble compounds (e.g. PCBs and chlorohydrocarbon pesticides) may be excreted in milk, and exert significant biological effects upon the young through breast feeding.

Carcinogenesis, mutagenesis and teratogenesis

Chemicals that can cause cancer (carcinogens), mutation (mutagens) or malformation of embryos (teratogens) have received special attention because these processes (carcinogenesis, mutagenesis and teratogenesis) can threaten the survival of individuals, entire populations and species.

Carcinogens are chemicals which cause tumours. There are two main types of tumour (neoplasm). *Benign* lesions involve proliferation of cell growth. The cells formed resemble the cells from which they originated but do not invade normal tissue or spread to other tissues. *Cancers* on the other hand are an irreversible conversion of normal cells to malignant cells, which depart from the normal features of the cells from which they originated and grow in an uncontrolled manner. At a later stage, these malignant cells can destroy nearby tissues, and migrate and invade other tissues (metastasis), and the target cells are somatic cells.

Mutagens are chemicals that cause heritable mutation (alteration of DNA and chromosomes).

Teratogens are chemicals that act on and cause malformation of embryos, resulting in congenital defects.

Until recently, carcinogenesis, mutagenesis and teratogenesis have been almost exclusively studied in human and mammalian systems. However, high incidences of neoplasms have been reported for feral fish in contaminated areas (e.g. Great Lake, Boston Harbour). The ecological implications and significance of these pollution-induced alterations are poorly known, especially for lower vertebrates and invertebrates.

It should be noted that carcinogens, mutagens and teratogens may not have a high acute toxicity (i.e. low LD_{50} values). However, their chronic, sublethal effects may be much more damaging than acute effects because of the long latent period (up to 20–30 years) before symptoms are expressed. By the time the effects have been discerned, it is already too late to contain the damage and a large population could have been exposed to the chemical for a long time and consequently, a large number of individuals could be at risk.

There is growing evidence of a strong correlation between mutagens and carcinogens. Of 175 known carcinogens, 90% are also mutagens, and of 108 noncarcinogens, only 13% are mutagens. Some examples of carcinogens are shown in Table 4.2.

The initiation and formation process for cancers is shown in Fig. 4.3. Most mutagens and carcinogens are strong electrophilic, biologically alkylating or acylating agents. These chemicals may contain highly stressed heterocyclic three or four carbon member rings such as epoxides, episulphides or lactones which have a strong tendency to nucleophilic ring opening. Some other

Table 4.2 Sources and normal entry routes of some carcinogens. The types of cancer caused by these carcinogens are also shown.

Chemical	Cancer type	Typical source	Normal route of entry
Aflatoxin	Liver	Mouldy animal feed or grain	Oral
Asbestos	Lung, pleura	Building materials	Inhalation, skin, oral
Vinyl chloride	Liver	PVC manufacture	Inhalation
Nitroamines	Nasopharyngeal	Salt fish, rubber additives, tobacco smoke	Oral
Benzo[a]pyrene	Skin	Tobacco smoke, incomplete combustion, smoked food	Oral
Metal dust	Lung	Industry, pharmaceuticals	Inhalation

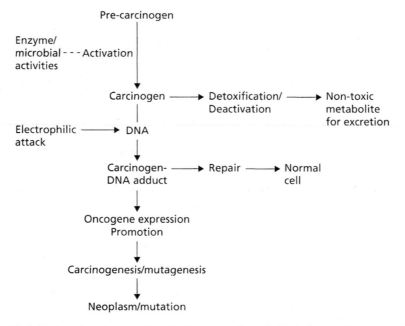

Fig. 4.3 Initiation of carcinogenesis and activation and metabolism of precarcinogens and carcinogens.

mutagens and carcinogens may produce oxygen or hydroxy radicals such as OH, H_2O_2 and singlet oxygen. These free radicals are highly reactive and may attack DNA causing strand breakage or damage to the purine and pyrimidine bases. A large variety of environmental carcinogens may be metabolically activated to become electrophilic metabolites which can attack nucleic acid and protein, and form covalent adducts. The formation of DNA–carcinogen

Table 4.3 Common examples of some carcinogen and mutagen groups.

Chemicals	Examples
Free radical-producing agents	Azo compounds, quinones
Polycyclic aromatic hydrocarbons	Benzo[a]pyrene 7,12-Dimethylbenz[a]anthracene
Halogenated aliphatic compounds	Vinyl chloride, trichloroethylene
Alkylating agents	Dimethyl-nitroamine, cyclophosphamide, nitrogen mustard
Acylating agents	Acid halids
Carcinogenic metals and metalloids	Cadmium, chromium
'Foreign bodies' that can disrupt intercellular homeostasis or mechanically interfere with DNA or chromatin	Asbestos, hard silicates
Compounds with similar structure as DNA bases (base analogues)	5-Bromouracil, 2-aminopurine

adducts is thought to be an essential step in carcinogenesis. A variety of organic compounds (e.g. quinones, azo compounds and aromatic nitro compounds) can generate highly reactive free radicals which can readily attack DNA and form adducts, leading to strand scission and chromosome breakage. The examples of some known chemical carcinogens and mutagens are given in Table 4.3.

The maintenance of DNA integrity is obviously of utmost importance to the survival and propagation of a species. The natural mutation rate is very low (estimated at one mutation per gene per 200 000 years). Under normal circumstances, DNA damage may occur as a result of wear and tear, random thermal collisions, ionizing radiation or ultraviolet light, attack by free radicals and actions of chemicals. DNA bases may be altered or lost and strands may be cross-linked covalently even during normal replication processes. These structural alterations may occur continuously under normal cellular conditions as well as during DNA replication.

In the course of evolution, cells have evolved sophisticated proofreading and repairing mechanisms to cope with normal wear and tear and to ensure that genetic information can be passed on to the next generation accurately and completely during DNA replication. In fact, a large proportion of the damage and lesions is continually being detected and repaired.

Because of this DNA repairing mechanism, most DNA adducts and lesions caused by mutagens and carcinogens will normally be repaired and may not necessarily lead to any observable effect. This is particularly true when levels of mutagens is low and excision and repair enzymes are not saturated by damaged DNA sites. Also, there is a long time lag between DNA damage and the occurrence of an observable result. These two factors obscure the interrelationship between cause and effect.

Field studies of carcinogenesis

Massive liver cancer appeared in the domesticated rainbow trout *Salmo mykiss* in fish hatcheries in the US in the 1960s. An extensive study showed that the high incidence of liver cancer was caused by contamination by aflatoxin in the feed ingredients.

Results of laboratory studies have shown that PAHs in Great Lake sediment can cause neoplasms of liver and skin in fish (the brown bullhead *Ameriurus nebulosus* and the white sucker *Catostomus commersoni*). Field studies showed that incidences of skin and cancer tumours in bullheads occurred mainly in severely contaminated locations. A 99% decline in surficial PAH during 1981–1987 was associated with a 75% decline in liver carcinoma in the brown bullhead during the same period. This evidence suggests that the incidence of liver tumour in the brown bullhead was directly linked to environmental carcinogens. The prevalence of epidermal papillomas in the white sucker in polluted areas also exceeded 60% whilst that in uncontaminated areas was only 5%.

In Sweden, an increased prevalence of skeletal deformity in the skull of the northern pike *Esox lucius* has been linked to three pulp mills in the Gulf of Bothnia. In a review of 41 regions in North America, clusters of liver and gastrointestinal tract and epidermal neoplasms occurring in feral fish (and to a lesser extent, invertebrates) were found to strongly correlate with contamination with known carcinogens.

A comprehensive study was carried out in the Pudget Sound. The prevalence of hepatic neoplastic lesions in the English sole *Parophrys vetulus* was related to both the incidence of DNA adducts and levels of PAHs and PCBs in the Pudget Sound sediment. Uptake and metabolism of PAHs by the sole was also demonstrated, and dosing with benzo[a]pyrene (B[a]P) resulted in the formation of B[a]P–DNA adducts and precancerous lesions in the sole. A similar correlation has been demonstrated between the high incidence of hepatocellular carcinomas in the winter flounder *Pseudopleuronectes americanus* and PAHs and polychlorinated hydrocarbon-contaminated sediment in Boston Harbour.

Teratogens

Most teratogens have a clear threshold below which congenital malformation will not occur. It is generally accepted that a significant concentration of teratogen is required to cause teratogenesis. Through the action of teratogenic chemicals, teratogens may cause selective cell death, altered biosynthesis of important metabolites or slowed growth, thereby leading to congenital

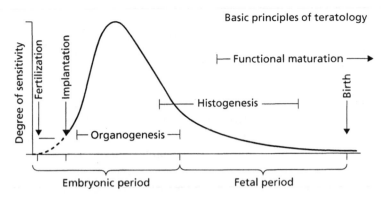

Fig. 4.4 Developmental stages and their relative sensitivity to structural malformations.

defects. Death of the embryo or fetus will lead to abortion. Susceptibility to teratogens is highly dependent upon the stage of development, and there are clearly 'critical periods' of teratogenesis at different developmental stages. Consequently, a teratogen administered at one developmental stage may lead to major congenital malformation whereas the same dose may have little or no effect when administered at another developmental stage (Fig. 4.4). Embryonic development may be conveniently classified into three stages: (a) implantation stage; (b) organogenesis stage; and (c) fetal stage. During the implantation/postimplantation stage, embryos have only a few cells, and any significant chemical disruption at this stage may seriously affect the viability of the embryo during its later development. The rapid cell division and cell differentiation occurring during organogenesis make this stage the most susceptible to teratogens. Exposure to teratogens during this developmental stage normally leads to major congenital structural changes in subsequent organ development (e.g. no limbs, heart defects). During the fetal stage, growth and functional differentiation of organs occurs, whereas most organogenesis has been completed. Exposure to teratogens at this stage may cause more subtle functional deficits such as growth or mental retardation.

Similarly to carcinogenesis and mutagenesis, teratogenesis has been well studied in humans but less well studied in other vertebrates and invertebrates. Many organochlorine and organophosphate pesticides, PCBs and selenium have been shown to be potent teratogens in birds. Retinoic acid is a highly potent teratogen for vertebrates. Malachite green oxylate is a fungicide commonly used as a therapeutic agent to treat fungal diseases in fish farming. It has been shown that this substance causes a three- to five-fold increase in skull and spine deformities and malformation of embryos when administered to eggs of the rainbow trout *Salmo mykiss*. Malachite green also cause fetal toxicity and a two- to three-fold increase in skeletal, liver, heart and kidney deformities in the fetuses of the rabbit *Oryctolagus cuniculus*. Deformities have been reported in many avian species from the Laurentian Great Lake, Canada. The overall results of research in this area provides evidence for a cause–effect relationship between the incidence of deformities in cormorants and terns and exposure to planar halogenated compounds, such as some of the dioxins, and total PCBs in the Great Lakes.

Likewise, a study on five populations of the snapping turtle *Chelydra serpentina* in Ontario indicated that deformities of tail, leg and jaws were significantly correlated with levels of PCB and organochlorines in the natural environment. PCBs have been shown to cause malformation of embryos in minks and other mammals. In seals, for example, reproductive failure, abnormal development of the reproductive system in females and abortion have been related to an elevated body burden of PCBs, hexachlorobenzene and dieldrin in the blubber. Several studies have related malformation of fish embryos and bone deformities in fish to levels of industrial pollution. It has been shown that pulp-mill effluents cause deformities of the gill cover in the perch *Perca fluviatilis* and upward bending of the jaws in the northern pike *Esox lucius*. The herbicide trifluralin has been shown to cause vertebral deformity in 10 species of marine fish, making them more vulnerable to predation and reducing their survival rate in their natural environment.

Testing and evaluation procedures

The Ames test developed by Bruce Ames is probably the most widely used test for identifying mutagens and carcinogens. The wild-type bacterium *Salmonella typhimurium* can synthesize histidine from glucose and a nitrogen source, and therefore does not require histidine in culture medium for growth and is termed his+. A mutant strain of the same species (his– mutant) has lost a single enzyme step in the biosynthesis of histidine and hence the ability to synthesize its own histidine. This mutant therefore cannot grow on culture medium without histidine. Mutagens and carcinogens may cause reverse mutation (change from mutant back to wild type) of the his– mutants to the wild type which can grow on histidine-free culture medium. Consequently, mutagenic and carcinogenic chemicals can be identified if the test chemical can cause a significant increase in the back mutation rate of the his– mutant. In the test, his– mutants are exposed to the chemical. The bacterium is then grown on a histidine-free medium. Bacterium colonies appearing on the histidine medium represent back mutation from his– to his+. A significant increase in the number of his+ bacterium colonies on the histidine-free medium compared with the spontaneous back mutation rate indicates that the chemical has caused back mutation and is likely to be a carcinogen or mutagen. However, since the bacteria cells are directly exposed to the mutagen or carcinogen in the test, while some chemicals may only become mutagenic or carcinogenic after biotransformation by the MFO system in the liver, the Ames test can only serve as a quick screening test for mutagens and carcinogens. Often, the microsomal fraction of rat liver containing the MFO enzyme system (known as the S-9 fraction) is incubated with the test chemical before exposure to the his– bacteria, in order to simulate xenobiotic metabolism in liver and detect precarcinogens and premutagens.

Obviously, whole animal tests are a much more reliable detector of mutagens and carcinogens. First, toxic effects are expressed at the organismal level. Second, they simulate more closely the route of entry of the chemicals. Third, biotransformation has also taken place so that precarcinogens and premutants can also be detected. The cost of whole animal tests, however, is very high and long experimental times are required. Also, the fact that a large sample is required to detect occurrence of cancers in the population prevents the use of whole animals as routine tests for carcinogens and mutagens. For example, it has been demonstrated that 161 animals need to be tested to provide a valid statistical sample size to detect a 10% increase in tumour incidence when the spontaneous occurrence rate is 1%.

Mutagens and carcinogens may also be identified by population epidemiological surveys. It has been well established that exposure to certain chemicals is closely associated with certain types of cancer. For example, the relationship between smoking and lung cancer is well established, and epidemiological surveys have also shown that the incidence of liver cancer in demersal fish is closely related to levels of PAHs in the sediment. Although epidemiological surveys can be a suitable way of identifying mutagens and carcinogens, this method is retrospective and cannot be used to regulate the use of new chemicals. Furthermore, epidemiological surveys are usually expensive and can only be used to detect chemicals that affect a relatively large percentage (> 10%) of the population. Exposure levels and interacting factors are also difficult to take into account in epidemiological surveys. Often, it is difficult to pinpoint the exact causative chemical and establish the cause–effect relationship, because typically there would be more than one common factor shared by the identified cohort.

Biochemical responses

Stress proteins

It has long been known that a class of low molecular weight proteins (ranging from 16 to 90 kDa) is inducible by heat shock stress. This class of heat shock protein (Hsp) is highly conservative and known to exist in all animal and plant groups studied. According to their molecular weight, these proteins have been classified into Hsp 90 (90 kDa) Hsp 70, Hsp 60 and Hsp 16–24. Numerous more recent studies have shown that their induction is not specific to heat shock, but also responsive to a variety of other stresses such as ultraviolet radiation, salinity changes, metals, toxic organic compounds and other xenobiotics at low, realistic environmental concentrations (and hence they have been renamed as stress proteins). For example, the low molecular weight

(20–30 kDa) stress protein is inducible by a variety of carcinogens, mutagens and teratogens. Exposure to heat shock and sodium arsenite causes rapid synthesis of Hsp 70 in the rainbow trout, and continuous synthesis of the protein may occur upon prolonged exposure to arsenite. Under normal conditions, these stress proteins occur at very low levels in cytoplasm, mitochondria and endoplasmic reticulum. The normal function of stress protein involves binding to target protein for modulation of protein folding, protein transport and protein repair. Another small (7 kDa) molecular weight protein, the ubiquitin, has been shown to be involved in DNA repair and replication. Under stress conditions, stress proteins are produced to protect the organism from damage to stress and to repair denatured proteins.

Enzyme inhibition and competition

A classic example of enzyme inhibition by toxicants is illustrated by the case of DDT. DDT was widely used as a pesticide to control malaria. This chlorohydrocarbon pesticide persists in the environment and has a very long environmental half-life. Because of its lipophilic nature, DDT has a high bioaccumulation potential and tends to concentrate in predatory birds. In many cases, the shell thickness, weight and total calcium content of eggs is significantly reduced and shell thickness shows a significant, inverse relationship to the body burden of DDT in the parent birds. The reduction in shell thickness caused a significantly higher incidence of egg breakage and hence a decline in the pelican population. Subsequent studies revealed that accumulation of DDE affected normal calcium transport and metabolism through inhibition of microsomal Ca^{2+}-ATPase in eggshell gland epithelium. The impairment of calcium metabolism reduced the shell thickness of the eggs.

Signal transmission in the nervous systems of animals involves both electrical transmission along the surface of the axon and chemical transmission of impulses between neurones. In all animals from annelids to mammals, acetylcholine (ACh) is the major chemical transmitter between neurones. ACh is discharged at a nerve synapse, moves across the synapse and binds to the ACh receptor in the postsynaptic membrane. The binding initiates an electric impulse in the next neurone and the 'message' then passes on. Termination of the signal transmission occurs when the chemical transmitter ACh is rapidly hydrolysed into acetate$^-$ and choline$^+$ by the enzyme acetylcholinesterase (AChE) released from the postsynaptic membrane immediately after the signal transmission. Organophosphates, carbamate pesticides and nerve gases used for chemical warfare inhibit acetylcholinesterase activity by binding irreversibly to AChE. Inhibition of chlolinesterase is mediated through blocking of the active sites of the enzyme that binds the ester group of the acetylcholine molecule. The binding removes functional AChE molecules, thereby causing an accumulation of acetylcholine at the nerve synapses and a continuous stimulation of nerves and their target muscles (Fig. 4.5). Cholinergic nerves

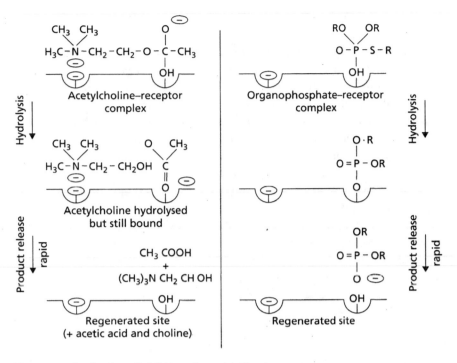

Fig. 4.5 Mode of action of inhibition of acetylcholinesterase.

include both voluntary motor nerves and involuntary nerves controlling the gastrointestinal tract. Continuous stimulation of the nervous system will therefore lead to continuous peristalsis of these voluntary and involuntary muscles, causing vomiting, respiratory distress, tremors and convulsions. As a consequence, symptoms of organophosphate intoxication typically include tremor and loss of coordination. Since AChE is such an important chemical, a large, excessive amount of AChE is available in the nervous system to ensure normal neural transmission function. Thus, in mammals, noticeable toxic effects will be expressed only when 50% of AChE is inhibited, and a 70% reduction is required before clinical illness is observable. Death will only occur when 80–90% of AChE activity is inhibited. Fortunately, many of the cholinesterase inhibitors can be metabolized or excreted quickly and will not accumulate. Also, ACh can also be regenerated fairly quickly and clinical recovery will occur.

Physiological responses

Respiration and photosynthesis

When molecular and biochemical responses culminate and are not compensated for by homeostasis, toxicants may begin to affect the normal physiological functions of an organism. A change in respiration rate is one of the common physiological responses to toxicants, and is easily detectable through changes in oxygen consumption rate. Numerous studies have shown that animals and plants may either increase or decrease their respiration (oxygen consumption) rate in response to a variety of toxicants such as metals, phenol and pesticides. Two possible strategies may be used by a living organism to cope with the impact of a toxicant. The first strategy may be to get rid of the toxicant through excretion, sequestration of the toxicant in inactive tissues (e.g. shell, fur, hair, skeleton, moult), or moving away from the area containing the toxicant. As discussed above, synthesis of specific proteins (e.g. MFO, metallothionein, stress proteins) may be initiated or increased to detoxify the toxicant, to sequestrate the toxicant in inactive tissues, to transform the toxicant to a more excretable form, or to repair damage. The increase in protein synthesis in the above processes is expected to be accompanied by an increase in respiration. Additional energy expenditure is also required for the metabolism, excretion and deposition of the toxicant, physiological compensation and avoidance behaviours. The increase in protein synthesis and energy expenditure will eventually be reflected in an increase in respiration (oxygen consumption) rate.

Many fish are able to swim away and avoid pollutants. Also to reduce exposure they may enter a dormant stage or reduce normal activities. For example: many bacteria and microscopic plants may form cysts; mussels may reduce feeding and close their valves to reduce uptake of toxicants. If normal physiological and metabolic functions are impaired by the toxicant, the respiration rate may also be reduced. For example, a reduction in the respiration rate of soil microbial communities when treated with toxicants has been reported. Changes in respiration (oxygen consumption) rate are reasonably easy to measure, and serve as a good indication of stress and intoxication of an organism. It should be remembered that these changes in respiratory response are not specific to any one toxicant.

In plants, photosynthesis may be affected by toxicants. The herbicides, atrazine and simeton, exert their toxic action on plants through inhibition of photosynthesis by preventing electron transport between photosystem I and photosystem II. Paraquat and diquat compete for electrons with ferrodoxin at the reducing end of photosystem I and divert electron flow and photosynthetic energy.

Feeding

Feeding is one of the most important physiological processes upon which depend the health and wellbeing, respiration, growth, fitness and even survival of a species. A reduction in the feeding rate of a variety of animals when exposed to toxicants such as lindane has been well documented. The feeding of many animals in their natural habitats (especially higher vertebrates like birds and mammals) depends on foraging success, which consists of a series of complex processes including locating, capturing and handling of prey. This series of complex processes is highly dependent upon chemosensory and visual mechanisms, prey recognition, nervous–muscular coordination and locomotion. Toxicants affecting any of these important physiological and behavioural functions may greatly reduce the foraging success and hence feeding of the affected species. For example, the foraging attempts and successes of fish are known to be reduced upon exposure to toxicants like heavy metals. Increases in turbidity due to sand dredging may adversely affect the foraging behaviour of the juvenile chinook salmon *Oncorhynchus tshawytscha* and the striped bass *Morone saxatillis.*

Growth and scope for growth

Growth is one of the important physiological parameters that can be directly related to an organism's ability to survive and reproduce. Growth represents the integration of feeding, assimilation, and energy expenditure over a relatively long period of time. Since a metabolic commitment is required to deal with intoxication and detoxification and feeding and food assimilation may be reduced by toxicants, the energy available for growth should be reduced if an animal is under toxic stress. Growth therefore serves as a good time-integrated indicator of the 'wellbeing' and performance of an organism over the medium to long term. Poor growth is likely to affect the time taken to reach maturity and possibly reproductive potential, which can directly affect the survival and fitness of individuals and populations in their natural habitat. One major problem, however, is that discerning differences in growth response may take a relatively long time, especially for a slow-growing species. Also, growth may also be affected by factors which may not be related to stress at all. In many invertebrates, for example, a significant amount of energy is channelled into reproduction, and growth is significantly decreased or negative growth may occur during reproductive seasons. The herbicide glyphosate inhibits plant growth by inhibiting the synthesis of aromatic amino acids and hence protein synthesis.

Growth may be measured directly or indirectly. Using the bioenergetic approach, it is possible to estimate the energy potentially available for growth and reproduction of an organism (termed *scope for growth*), as defined by:

scope for growth (SfG) = consumption − egestion − respiration − excretion

All bioenergetic items in the above equation are expressed in energy units (joules/unit time). The potential availability of energy for growth can reflect disruption of any of the important physiological processes mentioned above and balances the above energy budget equation. Compared with actual growth measurement, SfG is easy to determine and can be measured within a relatively short period of time. This approach is particularly useful for animals with slow growth rates. Food consumption can be measured by a simple feeding experiment, in which the difference in the wet or dry weight of food before and after a period of feeding is determined, and converted into energy-equivalent value (by calorimetry or proximate analysis). Egestion, respiration and excretion can also be measured experimentally.

Scope for growth represents the maximum amount of energy that can be spared for somatic growth. Thus, SfG provides a good indication of the 'ecological fitness' of the animal in the environment, and in the longer term, the chance of survival of the species. A good correlation between actual somatic growth and SfG has been demonstrated in a number of animals in laboratory and field studies, especially in bivalves. Reduction in SfG has been demonstrated in the bivalve *Venus verrucosa* after long-term exposure to petroleum hydrocarbons, and the reduction was associated with reduction in somatic growth and the general condition of the organism. Oxygen and salinity stress causes a reduction in SfG in the mussel *Mytilus edulis* and the reduction is correlated with reduced somatic growth and fecundity (as indicated by a reduced energy content of gametes). Reduced SfG in *Mytilus edulis* after exposure to diesel oil was correlated with a reduction in mass of gametes.

Reproduction

Species survival ultimately depends on reproductive success and the quality of offspring. Measurement of these endpoints is of great ecotoxicological interest. Reproduction is the most critical stage in the life cycle of a species as it has a determining effect on survival of the species. Any toxicants that may exert an adverse effect on reproduction are of considerable importance for several reasons. First, reproduction of many species can only be carried out successfully under optimal conditions, and any suboptimal conditions such as toxic stress may cause reproductive failure. Second, reproductive gametes are usually more susceptible to toxicants than adults, this is especially true in organisms with external fertilization. It should be noted that sperms and eggs are single cells. Their relatively small size and larger surface-to-volume ratio greatly enhances uptake of toxicants. For the same concentration and exposure period, uptake of toxicants by sperms and eggs would be much higher than for their multicellular adults.

Toxicants may affect not only the number of gametes produced, but also the quality of sperm or egg, clutch size or hatching rate. Exposure to sublethal concentrations of toxicants can cause a reduction in egg size, spawning ability and fertilization success.

Even if fertilization is successful, affected gametes may produce substandard offspring which have a lower chance of survival. Survival of some fish fry at their early developmental stages before feeding, for example, depends on the amount of yolk in the fish embryo. Fish fry with a smaller yolk are known to have a significantly lower survival rate than that of normal fry, probably due to the depletion of egg yolk before they can feed. Certain morphological defects caused by toxicants may also greatly reduce the fitness and hence chance of survival of the juvenile fish, which in turn, may lead to serious population effects. For example, a higher percentage of fish fry with skeletal deformities was found from industrial contaminated areas in the Great Lakes, and the survival rate of such deformed juveniles was found to be substantially lower than that of normal individuals, probably due to an increase in susceptibility to their natural predators. Perhaps one of the best examples illustrating the importance of egg quality on species survival may be illustrated by the impairment of thickness of pelican eggshells by DDT, through its metabolite DDE, mentioned earlier in this chapter. DDT did not affect the total number of eggs in the clutch, but since calcium metabolism was impaired by DDT, affected pelicans produced eggs with significantly thinner shells, which were easily cracked during the hatching process. The poor egg quality greatly reduced the reproductive success of the pelicans and caused a population decline. In a similar study, it has been shown that the body burden of organochlorine affected the eggshell thickness of the peregrine falcon *Falco peregrinus* and also caused a population decline. A study was carried out to compare the reproductive success of two populations of the double-crested cormorants *Phalacrocorax auritus* in Lake Michigan in the US and Lake Winnipegosis in Canada, in relation to contamination levels in these two locations. The environmental contamination was some seven to eight times higher in the cormorant eggs collected from Lake Michigan than in those from Lake Winnipegosis. Correspondingly, the proportion of eggs hatching from Lake Michigan was only 59%, significantly lower than the 70% hatching rate from Lake Winnipegosis. A significantly higher percentage of deformed bills was also found in Lake Michigan.

Fish, birds, mammals and other higher vertebrates normally have sophisticated reproductive behaviours, such as courtship, display and nesting. Impairment of any of these important reproductive behaviours will lead to reproductive failure and have a significant effect on the population.

Endocrine disruption

Some chemicals, albeit occurring in trace amounts (10^{-9} or 10^{-12} g/g level) in the environment, may cause significant disturbance to the hormonal system of animals and lead to serious environmental consequences at both individual and population levels. These chemicals are called *endocrine disrupters* (or sometimes environmental oestrogen because many of them affect normal oestrogen metabolism). In recent years, there has been growing concern about the hormonal disturbance caused by endocrine disrupters, mainly because these chemicals can cause a significant population effect at extremely low environmental concentration. Increasing evidence from laboratory and field studies has shown that trace amounts of many chlorinated hydrocarbons (e.g. PCBs), organophosphate pesticides, antifoulants (e.g. organotins) and diethyl-stilbestrol in the environment may cause significant endocrine disruption and reproductive disturbance and failure in invertebrates, fish, birds, reptiles and mammals. For example, laboratory studies have shown that chronic exposure to 0.5 mg L^{-1} diethylstilboestrol (DES) for 48 h alters steroid hormone metabolism in the water flea *Daphnia magna* and reduces its fecundity in the second generation.

Behavioural responses

An important sublethal effect of toxicants is the alteration of normal behaviour. Behaviours such as reproductive behaviours, foraging and feeding behaviours, predator avoidance, learning, orientation to a wide range of environmental factors, communication, social interaction and locomotion are often vital to the survival of a species in its natural habitat. Consequently, impairment of some of these behaviours may threaten the survival of the species in its natural environment.

To increase the chance of the continued existence of the species, all bisexual animals and plants have evolved mechanisms to bring their male and female gametes together. In lower plants and animals using external fertilization (e.g. urchins, clams and macroalgae), release of sperm and eggs into the environment is often synchronized. This is often achieved through chemical stimulation by the presence of opposite gametes in the water. Discharge of female gametes in marine animals (e.g. oysters and mussels) is often stimulated by the presence of male gametes of the same species in the water, and this mechanism serves to synchronize reproduction of the species to maximize fertilization success. In many species, the released sperms also locate and identify the eggs of the same species by chemosensory recognition. In other animals (e.g. insects), pheromones may be secreted to attract the opposite sex for mating, and also for identification and location of the same species. In higher animals (e.g. birds and mammals), mating will not occur until the male and female have gone through a series of complex and elaborate sexual behaviours, which may involve singing, courtship and display. This kind of reproductive behaviour has evolved to initiate, synchronize and ensure physiological and psychological readiness for sexual reproduction.

In the course of evolution, many predators have evolved sophisticated behaviour to maximize their foraging efficiency. Similarly, prey have coevolved behaviours to avoid their natural predators in order to minimize mortality. Many animals (such as mussels and oysters) may also exhibit rhythms in their feeding, so that their feeding activities can be synchronized with tides or other environmental factors. Toxicants may affect feeding and foraging behaviours, resulting in a cessation of or reduction in feeding, a change in the number of prey attacked and captured, a prolongation of prey handling time and reduction in foraging efficiency, and affect predator–prey relationships in the natural environment. Alteration of such behaviour may reduce foraging efficiency and hence growth and survival of the population. For example, the prey-capturing ability of mummichogs (*Fundulus heteroclitus*) from a mercury-polluted tidal creek was significantly lower than that of their conspecifics from a unpolluted stream. Sublethal concentrations of toxicants may also alter some of the important normal behaviours of the species, making them more vulnerable to their natural predators and reducing their chance of survival. It has been demonstrated that clam burrowing time (ET_{50}) had a direct relationship to pore water copper concentration in sediment.

Conclusion

Toxicants must be taken up from the environment into biological systems and transported to receptor sites before toxic action can occur. Chemicals taken up must pass through a biological membrane, and the uptake rate will depend on the chemical nature of the compound. Mixed function oxidase is the principal enzyme system involved in transformation and detoxification, and is inducible by toxicants. Transformation of xenobiotics normally involves turning the molecule into a more polar, hydrophilic compound to facilitate excretion. In Phase I transformation, a polar group will be added to the toxicant molecule. In Phase II transformation, endogenous metabolites may be conjugated to the Phase I metabolic products to further increase the polarity. Toxicants may be excreted out of the body, or sequestrated in inactive tissues to reduce toxicity. A number of specific proteins such as metallothioncin and heat shock (or stress) proteins may be synthesized to sequestrate toxicants or repair damage.

Genotoxicity may involve substitution, addition, deletion of DNA bases, frame shift mutation, adduct formation, intercalation of DNA, chromosome aberration, strand breakage or sister chromatid exchange. Most mutagens and carcinogens are strongly electrophilic, biologically alkylating or acylating agents, or carry free radicals enabling them to react with DNA. Some chemicals may not be mutagenic or carcinogenic in themselves, but are activated into mutagens and carcinogens by normal biotransformation. Most DNA adducts and lesions caused by mutagens and carcinogens will be repaired by cells and genotoxicity only occurs when the repairing mechanisms have failed.

Teratogens may lead to fetal death or malformation of embryos. Susceptibility to teratogens is dependent upon developmental stages, and there are clearly 'critical periods' of teratogenesis at different developmental stages.

Results of several field studies clearly demonstrate a close relationship between levels of environmental contamination, body burden of toxicants, and incidences of cancers, mutation and malformation of embryos in feral invertebrate, fish and bird populations. Effects of mutagens and carcinogens on natural populations are not clearly evident, while clear examples can be cited to support population decline caused by teratogens.

Toxic effects may be caused by disruptions to normal biochemical functions through competitive and noncompetitive enzyme inhibition. Toxicants may also exert their toxicity by competing with normal metabolites or essential elements for binding sites, thereby interrupting their normal function. Patterns of energy dislocation are sensitive to toxicant exposure, although this biochemical response is nonspecific to toxicants.

Common physiological responses to toxicants include changes in respiration or photosynthetic rate, feeding, growth, scope for growth and reproductive performance. Both laboratory and field studies have clearly demonstrated that certain chemicals such as pesticides and PCBs may cause significant reproductive dysfunction in animals at very low concentrations. These endocrine disrupters are of growing concern.

Certain behaviours involved in reproduction, feeding, orientation and social communication are essential to the survival of species in their natural environments. There is good laboratory and field evidence to suggest that these important behaviours may be impaired by sublethal concentrations of toxicants.

Toxicant-induced responses may occur at different biological organizational levels. Responses at lower biological levels may be compensated for by homeostasis, or else subsequently translated into biological effects at higher organizational levels. Some of the sublethal responses may be reversible while others may be irreversible. Some responses may lead to significant biological effects while others may not. Many of these biochemical, physiological and behavioural responses may have long-term effects on natural populations. In general, chemicals impairing fitness traits (e.g. growth, reproduction, recruitment and alteration of the gene pool) are likely to pose a significant threat to the survival of the species. The toxic effects of these chemicals should therefore be given serious attention in environmental protection and management.

Further reading

Francis, B.M. (1994) *Toxic Substances in the Environment.* John Wiley & Sons, New York.
Peakall, D. (1992) *Animal Biomarkers as Pollution Indicators.* Chapman & Hall, London.
Rand, G.M. (1995) *Fundamentals of Aquatic Toxicology. Effects, Environmental Fate, and Risk Assessment,* 2nd edn. Taylor & Francis, Washington.

5: Effects on Population, Community and Ecosystem

Introduction

Generally speaking, suborganismal responses, as described in Chapter 4, are sensitive to toxicants, relatively easy to determine, detectable within a short period of time, and indicative of early stages of toxicant stress. Thus, many of these responses are being used as an early warning system of environmental degradation before significant biological effects become observable. However, although many of these responses may reflect the health condition of individuals, they may not necessarily lead to adverse ecological consequences. It should be noted that there are fundamental differences between human toxicology and ecotoxicology: the major concern of human toxicology is with the individual. Thus, much of human toxicology work has been focused on suborganismal (e.g. biochemical and physiological) responses, with the aim of protecting individuals. On the other hand, the fundamental tenet of ecotoxicology is to protect not individuals, but populations, communities and ecosystems in the natural environment. As such, any observed suborganismal responses must be ecologically relevant, and should be related to adverse ecological consequences, otherwise they will be of limited value in environmental management. As opposed to protecting human individuals, it might be argued that toxicants affecting (or eliminating) a few individuals in the natural environment and yet failing to exert a significant ecological consequence on populations, communities or ecosystems could be deemed to be 'acceptable'. To support this argument and clearly illustrate this important concept, let us take fishing as an example. Fishing activities will remove a certain number of fish from the ecosystem, but properly managed fishing would not be expected to cause any significant effect on the ecosystem. Fishing is therefore considered to be ecologically acceptable. Overfishing, on the other hand, is ecologically unacceptable and of great environmental concern, since overfishing will have adverse ecological consequences on natural populations and communities such as extinction of populations and major changes in trophic structure, species composition and ecosystem function. This is an important concept, discussed in Chapter 1, which stressed that ecotoxicological studies and endpoints should ultimately be related to ecological changes in the receiving environment. This also explains why ecotoxicological endpoints relating to Darwinian fitness traits (e.g. growth, reproduction and survival) are important, since these endpoints can be more easily related to adverse ecological consequences at population levels and above.

The ultimate goal of ecotoxicology, in its applications, is to protect the environment from the toxic effects of chemicals. In this respect, we may like to ask ourselves the following questions:

What exactly do we want to protect?

Do we want to protect every single species in the whole ecosystem?

Do we want to protect biodiversity?

Are we merely interested in protecting commercial species for our sustainable exploitation, or are we interested in protecting certain endangered species?

Are we only interested in protecting the integrity (structure and function) of community and ecosystem, or are we interested in protecting public health?

It is obvious that the focus and endpoint of ecotoxicological investigations will be different, depending on our priorities and targets for protection. In tropical habitats, about one million species have been described so far, and some 20 million species are suspected to exist. It would therefore be extremely difficult to protect *all* 20 million species in the tropical ecosystem with confidence, because we can direct our evaluations to only a few species. If we are going to protect *all* species with confidence, perhaps zero discharge of chemicals is the only answer. If we are only interested in protecting certain endangered species or commercial species in the ecosystem, we should be more interested in the toxic responses of these target species, and other species on which these target species depend. If public health is the key issue, studies and efforts should be devoted to commercial food species at higher trophic levels, to ensure that levels of contaminants in food harvested from the ecosystem will not exceed our acceptable public health limit. If we want to protect the integrity of community and ecosystem, our targets for protection should be those species which are most important in regulating the structure and function of ecosystems.

Obviously, there are great advantages in using population and community responses as ecotoxicological endpoints. The major problem involved, however, is that population and community responses are much more difficult to study and determine, as compared with the suborganismal responses discussed in Chapter 4, for the following reasons. First, the degree of biological complexity and variability increases as one moves up the ladder of biological organization (i.e. from the molecular to ecosystem level). Population, community and ecosystem changes are typically associated with a high degree of variability. The high background variations compared to signal–effect ratio in responses observed at higher biological organizational levels greatly reduce the precision and reproducibility of measurement, and increase correspondingly the degree of uncertainty in prediction. Because of this inherent limitation, only large differences in effects can be detected. Second, toxic effects will only be expressed at a higher level of biological organization when compensation fails at the lower levels. Various homeostasis and compensation mechanisms operating at the lower biological organizational levels tend to dampen the expression of toxic effects at higher organizational levels. Third, because of homeostasis at various levels, the expression of toxic effects at population and higher levels typically takes a relatively long period of time (e.g. from months to years). Significant ecological damage will have already been done by the time population responses become evident. Fourth, ecological interactions (competition, predation, trophic relationship, etc.) at population and community levels are typically complex and variable, and compensation mechanisms and recovery at the population level and above are much less well understood. Thus, it would be much more difficult to translate toxic effects observed in one species to indirect effects on other populations in the same community or ecosystem. Fifth, responses at higher organizational levels are normally much less specific to toxicants. For example, it would be more difficult to attribute an observed decline in a natural fish population to elevation of contaminants in sediment, water or some other phases of the environment. The above limitations make it difficult to assess ecological effects of toxicants at population levels and above, and these problems have become the major challenges to ecotoxicologists.

Influence of life-history stage

Most ecotoxicologists consider that a life-cycle test is the 'ultimate' test in arriving at the long-term 'safe' concentration. However, since a life-cycle test entails exposing the whole life cycle of the test organism (from embryo to embryo) to toxicants, the test normally requires some 6–12 months to complete and the high cost of such a test is inhibitory. Also, the life-cycle test is often not possible for many species because of the technical difficulties involved in keeping the animals in captivity under laboratory test conditions for a long period of time, let alone in reproductive condition.

It is well known that susceptibility, reproductive impairment and reduction in survival within a single species can vary drastically according to the age as well as the developmental stage of individuals. Experience from life-cycle tests on a variety of species and chemicals shows that certain developmental stages of an organism are consistently more sensitive to toxicants. The susceptibility of the water fern to toxicants has shown to change considerably depending on the life stage of the species. The larval and juvenile stages are generally much more susceptible to toxicants. Given the same exposure concentration and exposure period, uptake of toxicants by larvae and juveniles will be much higher, because of their small size and a much higher body surface–volume ratio. Furthermore, detoxification enzymes may not be well developed at early developmental stages in the life cycle. This makes the larvae and juveniles less able to cope with toxicants. In human beings, nitrite may combine with haemoglobin to form methaemoglobin, making the haemoglobin unable to take up oxygen. In adults, this situation is corrected by the enzyme methaemoglobin reductase, which will convert methaemoglobin back to haemoglobin. This enzyme is not produced in infancy, and infants drinking water with a high nitrate or nitrite content have a much lower capacity to take up oxygen in their blood, since their haemoglobin is bound by nitrite and converted into methaemoglobin. This causes a disease known as the *blue baby syndrome* because of the low level of oxygenated blood. Since early life stages are normally more sensitive, most life-cycle tests tend to concentrate on early developmental stages. For example, many pesticides are not designed to kill all ages or stages but only the most sensitive stage of the target species. This will reduce their level of usage and damage to other nontarget species. The reproductive stage is normally the most critical stage in the life cycle of a species, since reproductive success (i.e. the number and quality of offspring produced) is the ultimate determining factor for survival of a species. Thus, any impairment of reproductive success will lead to serious, long-term, ecological consequences at the population level, particularly if density-dependent compensation mechanisms are absent. For example, if a toxicant reduces the reproductive success of a species by 50%, the population of the species will be reduced to only 12.5% of its original size after only three generations. It should be noted that the probability of encountering the opposite sex for reproduction may be drastically reduced if the population density of a species falls below a certain limit. The reduction in mating opportunities may itself lead to population extinction.

An example of these factors has been observed with levels of polychlorinated biphenyls (PCBs), and TCDD equivalent have been shown to affect the reproductive success of the double-crested cormorant *Phalacrocorax auritus*. Levels of PCBs and dioxins which in eggs from Lake Michigan (7.8 μg g^{-1} and 138 pg g^{-1}, respectively) were seven to eight times greater in eggs from Lake Michigan, than those in Lake Winnipegosis (1.0 μg g^{-1} and 19 pg g^{-1}, respectively). Correspondingly, the hatching rate of eggs at Lake Michigan (59%) was significantly less than that at Lake Winnipegosis (70%). Also, 0.79% deformed bills were found in the Michigan population as compared to 0.06% in the Winnipegosis population. Similarly, the total level of PCBs in eggs of the Forster's tern *Sterna forsteri* from Lake Michigan in 1983 and 1988 has been related to reproductive success of the birds. A reduction of 67% of total PCB residues was found in the eggs between 1983 and 1988, and the reduction in toxicant level was accompanied by a significant improvement in reproductive performance during the same period. Another long-term study on herring gulls in the Great Lakes showed a dramatic increase in reproductive success from 0.2 fledglings/nest to 1.0 fledgling/nest between 1975 and 1977, and this increase in reproductive success was related to a rapid decline in concentration of halogenated aromatic hydrocarbons in eggs and attributed to the reduced toxicity of PCB, TCDD and polycyclic aromatic hydrocarbon (PAH) to the embryo.

The high levels of DDT, PCBs and organochlorines in the Baltic Seas in the mid-1950s markedly reduced the hatching rates of eggs (from 72% to 25%) and the nesting success (from an average of 1.8 to 1.2 fledging/nest) of the white-tailed eagle *Haliaeetus albicilla* in the 1960s and 1970s. The nesting and reproductive success showed a steady increase following the ban of DDT and PCBs, and the reproductive success of the white-tailed eagle population in 1994 almost resumed the values prior to the occurrence of organochlorines in the Baltic. It has taken some 15 years to remove the negative effects of DDT on eagle reproduction, and another 10 years for the population to recover. A similar picture has emerged in a parallel study on three species of seals (harbour, grey and ringed seals) in the Baltic. The high body burden of DDT and PCB led to skeletal and uterine deformations in the seals. The average percentage of seals with uterine deformation decreased from 36% in 1977–1986 to 25% in 1987–1993. The grey seal *Halichoerus grypus* population in the northern Baltic showed a marked increase since the banning of DDT and PCB in the Baltic region. A similar recovery was also found for the other two species.

It is not only reproduction which is important in determining species survival. Survival of the species also depends on the ability of the species to maintain the number of surviving reproductive individuals in the next generation. Thus, high mortality at the larval and juvenile stages may also lead to population decline.

Genetic variation is also another important factor contributing to the susceptibility of a species. Certain genotypes may be particularly susceptible, or particularly tolerant, to a given toxicant. Thus, the degree of genetic heterogeneity may also affect the response of a population to toxicants. For example, it has been shown that inbred Wistar rats give a steeper dose–response curve compared with genetically mixed animals. Similarly, it has been shown that the susceptibility of the water flea *Daphnia maga* (a 'standard test species' in ecotoxicology) varies considerably, depending on the genetic strain of the test animal. One dilemma facing ecotoxicologists is that in laboratory studies, test organisms of pure genetic strains are preferred and selected, in order to reduce experimental variations and improve the precision and reproducibility of the test. However, results derived from pure strains may be limited in terms of predictive power when extrapolating laboratory results to field populations with a much greater genetic diversity.

Influence at population level

In any multispecies assemblage in the natural environment, some species are liable to be more susceptible to a given toxicant, while others are more tolerant. Upon the introduction of a given toxicant, species which are particularly sensitive will be eliminated before the others, while the tolerant population may grow faster since extra resources are being spared by their competitors which have been eliminated by the toxicant. Thus, the absence or abundance of certain species is often indicative of the level of a toxicant in the environment. Perhaps one of the best known examples is the natural distribution of lichens. Lichens are symbiotic plants comprising a photosynthetic alga or cyanobacterium and fungi. Most species of lichen are very sensitive to moisture containing dissolved airborne pollutants. The survival and presence of lichen would therefore indicate a low level of airborne pollutants in the environment, and vice versa. Likewise, the polychaete worm *Capitella capitata* is particularly tolerant to organic pollution, and becomes abundant in marine sediment with high organic loadings. The occurrence and abundance of certain species, such as lichen and *Capitella capitata*, may therefore serve as a convenient indicator of the presence or level of a certain type of toxicant in the environment.

Low concentrations of toxicant may not necessarily eliminate a species, but only reduce its abundance in the environment by reducing the intrinsic growth rate and reproductive potential of the species. Toxicants may also affect recruitment and threaten species survival. For example, an oil spillage occurring during the recruitment season may reduce new recruits of barnacles and mussels and lead to a population decline in subsequent years. Toxicants may also selectively eliminate certain susceptible age groups in a population, thereby altering the population structure and hence population dynamics and species survival in the long term.

Alteration of gene pool and gene frequency

Toxicants may cause a change in the genotype frequency and gene pool of a population by eliminating certain susceptible genes or promoting certain resistant genes. One of the most commonly cited examples is the use of insecticide in eradicating agricultural pests. It has long been recognized that certain pesticides may become ineffective after prolonged use. This is because a few individuals within the sprayed population may be particularly tolerant to the insecticide applied. These few tolerant individuals are able to survive and reproduce after the pesticide application, and give rise to offspring carrying the resistant gene. Offspring carrying the 'resistant' gene soon monopolize the resources and habitats spared by the eliminated individuals that carried the 'susceptible' gene. Because of this artificial selection process, the gene pool in the surviving population will soon be dominated by the resistant strain. The change in the gene pool makes it necessary for farmers to change to another pesticide from time to time. Alteration of the gene pool by chemicals and toxicants has also been demonstrated in many other populations. One example is the change of gene pool in bacterium populations after the administration of antibiotics in marine fish farming. Oxytetracycline is an antibiotic commonly used as a feed ingredient to control fish disease. Marine sediments underneath culture cages are contaminated with oxytetracycline through deposition of feed wastage. It has been shown that levels of up to 100% of the population of oxytetracycline-resistant bacteria occur in marine sediment near fish farms after administration of this antibiotic as medication to diseased fish, and the resistance persisted for more than 13 months.

Alteration of ecological relationships

The effects of a toxicant affecting one species in the ecosystem may indirectly affect other species through a range of species interactions such as predator–prey relationships, trophodynamic relationships, exploitation or competition between species (two consumers exploiting the same resources) and mutualistic interactions. Studies carried out on the effects of toxicants on ecosystems of mammalian target species have shown that indirect effects may be as important as direct effects of toxicants on target species. For example, killing off insect pests may well destroy the food supply for other valuable predator species (e.g. salmon). While the insect populations may recover within months, the slow-breeding predators may take several years to recover. Changing one population may also affect other populations through altering ecological relationships such as competition and the predator–prey relationship. The wolf is one of the major predators of deer in the grassland habitats of North America. The elimination of wolves in North America has led to an explosion of the deer population. Consequently, the uncontrolled grazing by the deer population has caused great damage to forests and grasslands.

Studies on indirect effects are often very complicated and difficult, and an extensive knowledge of species interactions within the ecosystem is required. One point clearly illustrated from the few examples available is that our prediction of ecological consequences based on a single-species study or test may often be erroneous.

Impact on public health

It is well known that toxicants such as metals and organohalides may be bioconcentrated leading to elevated concentrations in some biota. These high levels may not only affect animals in the natural system, but also threaten human public health. In particular, filter feeding bivalves (e.g. mussels and oysters) have a remarkable capability to concentrate various kinds of toxicants from the environment with concentration increases up to a million-fold being quite common. These high levels of toxicants may pose a threat to public health and be detrimental to the shellfish resource. There is increasing evidence to show that levels of contaminants in fisheries resources in many places are approaching the existing public health limits.

Effects on community and ecosystem structure and functions

In ecology, two major types of community and ecosystem parameters may be distinguished and used as endpoints in ecotoxicological studies: structure and function. *Structure* of a community or ecosystem refers to the 'snapshot' properties (e.g. how many, what kind, what conditions), while *function* refers to

Table 5.1 Trends in function expected in stressed ecosystems.

Energetics
 1 Community respiration increases
 2 P/R ratio becomes unbalanced
 3 P/B and R/B ratios increase as energy is diverted from growth and reproduction into acclimation and compensation
 4 Importance of auxiliary energy increases (import becomes necessary)
 5 Export of primary productivity increases

Nutrient cycling
 6 Nutrient turnover increases
 7 One-way transport increases and internal cycling decreases
 8 Nutrient loss increases

Community dynamics
 9 Life spans decrease, turnover of organisms increases
10 Trophic dynamics shift, food chains shorten, functional diversity declines, proportion of energy flowing through grazer and decomposer food chains changes

General system-level trends
11 Efficiency of resource use decreases
12 Condition declines
13 Mechanisms and capacity for damping undesirable oscillations change

the dynamic properties and ecological processes over time (e.g. how fast, how quick, how well certain functions are being performed). A summary of trends expected in a stressed ecosystem is shown in Table 5.1.

Structure of community and ecosystem

Structure of a community is commonly described or quantified by parameters such as species number, absolute or relative abundance of species, species composition, species dominance, biomass, species richness, species diversity, trophic levels and similarity between communities. These community structural parameters may be selected as ecotoxicological endpoints to compare and evaluate changes in a particular community before and after the introduction of toxicants, or between 'control' and 'treatment' communities.

Measurement of abundance is often straightforward. Representative subsamples may be taken and measured in cases where organisms are numerically abundant. A large sample size is required if rare species are to be covered. Electronic counting and sensing devices are now available (e.g. Coulter counter), and may be used to facilitate the enumeration of organisms in a sample. Some species (e.g. lichen, grass, coral, macroalga) may be difficult to count, and percentage coverage may be estimated to indicate abundance. The recent advent of image analysis technology provides a quick method for estimating abundance. Photographs or video footage may be taken in the field or in the laboratory, and the image of each species can be differentiated and recognized by its colour, shape and size, then counted and recorded. Abundance estimation of some species (such as small mammals and birds) may necessitate the use of indirect methods, such as counting the number of faecal pellets, songs or footprints.

Determination of biomass is also relatively simple. Biomass is normally expressed in terms of wet weight, dry weight or ash-free weight (which indicates the amount of organic content in the sample).

Determination and comparison of species composition entail species identification, and normally requires considerable taxonomic expertise. Identification of certain taxonomic groups (e.g. nematodes, minor phyla, protozoa and bacteria) is known to be particularly difficult, and often calls upon input from specialists of a particular taxonomic group. Theoretically, a stressed community would carry a lower number of different species than its undisturbed counterpart in a similar habitat, since sensitive species may be eliminated in the former situation.

Species diversity is a function of both number of species as well as distribution and abundance within individual species. Ecosystems with a low diversity are less stable and the system is more vulnerable to perturbation. Based on this principle, species diversity has been used extensively in pollution studies, to indicate the 'health condition' of ecosystems. In theory, a toxicant would reduce the number of species in a community by eliminating certain sensitive species. At the same time, the abundance of a few tolerant, opportunistic species in the community will increase. Such a decrease in species number and increase in abundance of a few species will be reflected by a decrease in species diversity. Thus, the use of species diversity as an ecotoxicological endpoint for detecting toxic stress in a community has a sound theoretical basis. Various diversity indices have been developed, to provide a mathematical means of describing and comparing differences in species diversity between habitats, or changes temporally within a habitat. These indices are sensitive to changes in species number as well as in the distribution of individuals within species, and have been used extensively to detect community changes induced by pollutants or toxicants. Amongst these diversity indices, Shannon–Weiner's index (H'), Evenness (E), and Margalef's species richness (D) are those most commonly used.

Shannon–Weiner index (H') = $-\Sigma N_i/N \ln (N_i/N)$

Evenness (E) = $H'/\ln S$

Margalef's species richness (D) = $(S - 1)/\ln N$

where S is total number of species, N is total number of individuals and N_i is number of individuals in the ith species ($i = 1$ to S).

The major criticism regarding diversity indices is their insensitivity to change in a few rare species, and also to replacement of the same number of, but different species. Also, diversity indices are sample-size dependent, and therefore can only provide relative comparisons between control and polluted sites, or changes over time, provided the sampling environment, methodology and sampling equipment are identical.

Another technique that has been used to detect community changes is the log normal distribution. Log normal distribution is a graphical probability plot method used as an index to indicate changes in the abundance and distribution of species in a community. It would generally be expected that the sum of many independent events and factors would assume a normal distribution. Since the size of a population is affected by a large number of independent factors, it is expected that a sample of the population will also assume a normal distribution. The abundance of each species in an unperturbated (normal) community will tend to assume a normal distribution, in which most species in the community would be moderately abundant, while there are some rare species which would occur in low numbers. When a community is subjected to toxicant stress, certain species will tend to increase at the expense of others, thus causing a departure from the normal distributional pattern. Similar to species diversity, however, the normal distribution technique is affected not only by pollution or environmental disturbance of toxicants, but also by any other natural factors that alter natural processes (e.g. seasonal recruitment of juveniles or spats). Good background data are therefore required to differentiate natural and anthropogenic causes. Some workers have suggested that deviation from the normal distribution of individuals among species could provide a more sensitive indication of pollutant impact than species diversity.

In assessing the toxic responses of communities, it is of interest to compare the similarity of species composition between different communities, as well as before and after the introduction of toxic stress. A variety of *similarity indices* have been developed to provide quantitative comparison of species composition and abundance between communities, or the temporal changes within the same community. Simple similarity indices may be based only on presence and absence data (e.g. Jaccord's index and Sorenson's index) as illustrated below.

Jaccord's index $= (C/A + B + C) \times 100\%$

Sorenson's index $= 2C/(A + B + 2C) \times 100\%$

Coefficient of similarity $= [C \times 100/(A + B - C)] \times 100\%$

where C is number of species occurring in both communities, A is number of species found in community A only and B is number of species found in community B only.

Since the above indices are based on presence or absence data, they are only sensitive to the addition or removal of species and insensitive to change in a population where no species is lost or added. *Quantitative similarity indices*, which take into consideration species-weighted numbers, are also available.

This type of similarity index is sensitive to the loss or addition of species, as well as to changes in abundance of individuals within species. Ideally,

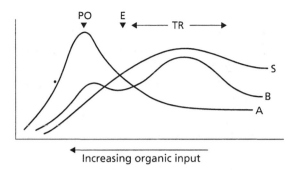

Fig. 5.1 Generalized spatial changes along a gradient of organic enrichment developed by Pearson & Rosenberg (1978).

community comparison should be sensitive to loss or addition of species, and also to changes in abundance in both low- and high-frequency taxa and also able to detect proportional changes. Nevertheless, none of the existing indices can fulfil all these desired criteria. Moreover, the answer to the question 'How similar is similar?' may be rather subjective. Some workers have suggested that > 50% similarity is considered to indicate a high affinity between samples while a value of 25% is arbitrarily taken as separating low and intermediate levels of affinity. Multivariate techniques (e.g. agglomerative, hierarchical classification) are also available to compare similarity of community between studies and sites.

Based on their studies of spatial changes in marine benthic communities along gradients of organic pollution, it has been concluded that there are several community responses to pollution stress. These community responses are quite specific, and yet also applicable to other toxic discharges. Benthic infaunal communities are found to be organized structurally, numerically and functionally in relation to the organic pollution gradient. The Pearson–Rosenberg model, illustrated in Fig. 5.1, classified spatial changes in soft-bottom macrofaunal communities along the pollution gradient into the following zones: (a) an azoic zone; (b) a poor community dominated by a large number of opportunistic species and a small peak of biomass; (c) an ecotone point characterized by low biomass, low abundance and high diversity; (d) a transitional community with maximum species number and biomass; and (e) a community with stable species number, biomass and abundance. Benthic animals were absent near the source of discharge, since none of the species can survive under extreme conditions. At a distance away from the discharge where stress is severe, a few opportunistic species able to survive the harsh conditions (e.g. the polychaete worm *Capitella capitata*) dominate the entire community. These opportunistic species are characterized by high abundance and biomass, because of the increase in population due to the reduced pressures of competition and predation, and also because they can exploit the resources spared by other species eliminated by pollution. Further away from the pollution source where level of stress is moderate, more species in the benthic community can withstand the low level of stress. A reduction in rarer species and an increase in the abundance of some of the moderately common species are therefore observed.

This generalized spatial pattern is generally supported by many other pollution (e.g. pulp-mill effluent) studies on benthic communities in the temperate and tropical sea lochs (semi-enclosed, low-energy depositional habitats), irrespective of the type of toxicant and stress involved. A study based on quarterly sampling of a benthic community over 4 years (1985–1989) near the Orange County outfall in California (an open oceanic environment with high-energy, erosional habitats) however, showed the following departures from the Pearson–Rosenberg model: (a) no sharp decline in species-abundance–biomass (SAB) curves to azoic conditions; (b) displacement of SAB curves away from the outfall; and (c) no exclusion or elimination by opportunistic species of rare species. The results of this study suggest that the SAB curves and the Pearson–Rosenberg model may have to be modified in high-energy habitats.

Key species

In maintenance of the community structure, not all species within the community are equally important. Some species play a key role in controlling the structure of the community; they are therefore called the *key species* or *keystone species* and are much more important than the other species in the same community. Should the key species be affected, drastic changes in the community structure may be expected. Limpets, for example, are the key species in the rocky shore habitat because they are important in controlling the structure of intertidal rocky shore communities. The grazing activity of limpets controls the density of macroalgae on rocky shores. Experimental removal of limpets will lead to overgrowth of macroalgae, and the high density of macroalgae will in turn, retard larval settlement and establishment of other species such as barnacles and mussels on the shore. Movement of limpets also shovels the newly settled spat of mussels and barnacles, thereby keeping the density of barnacles and mussels down. Removal of limpets would therefore result in major changes of community structure in the intertidal rocky shore. Buffalo is the key species in the North American prairies. The grazing activities of buffalo prevents the forest from invading the grassland. A reduction in the number of buffalo will reduce their grazing activities and encourage the encroachment of forest. This will consequently cause a major structural change in the community. From these two examples, therefore, it is clear that protection of the key species is of major importance if community structure is to be maintained.

Indicator species

A community is composed of an array of species which interact with each other, and many species will share the common resources such as food, water, refuge and space. There is considerable variability amongst species in their susceptibility to toxicants. Low concentrations of toxicant may first eliminate a few susceptible species in the community, and subsequently those species that solely rely on the susceptible species. After the elimination of the susceptible species, the resources originally utilized by these susceptible species may be spared for other tolerant species that share the same resources. Given extra resources spared by the susceptible species, some of the tolerant species may take advantage and increase their number ('opportunistic species'). Alternatively, species which cannot normally establish themselves in the community because their inferiority in competing with the susceptible species may now be able to do so in the absence of the susceptible species. Likewise, removal of a susceptible predator will encourage the growth of its tolerant prey. As a consequence, a number of these opportunistic species will increase rapidly at the expense of the susceptible species, and this will lead to a change in species composition and diversity in the community. A typical example is provided by the polychaete *Capitella capitata* discussed earlier. Because of its high tolerance to low oxygen and organic enrichment, this polychaete worm dominates in organically polluted sediment.

Functions of communities and ecosystems

In general, we have much less experience in the use of functional endpoints than in the use of structural endpoints in ecotoxicological studies. Important community and ecosystem functional endpoints in ecological studies include primary productivity, secondary productivity, nutrient flux rate, nutrient recycling rate (e.g. nitrification, denitrification, carbon cycle), energy flow rate, decomposition rate, community respiration, resiliency (the capacity of ecosystem to recover from stress), production/biomass ratio, production/respiration ratio and biomass/energy flow ratio.

Community respiration provides a measure of the amount of energy required to maintain the system, and is sensitive to toxicant stress. Community respiration is often estimated from oxygen consumption. Biochemical oxygen demand, for example, is a widely used measurement used by pollution scientists and engineers to determine the respiration rate of the microbial community in a water body. For measuring respiration in benthic communities, a variety of bottom respirometers have been developed.

The production/biomass (P/B) ratio has been suggested to be a useful indicator of ecosystem maturity. In a mature ecosystem, the P/B ratio is close to 1. A significant deviation from this value would indicate stress in the system. The ratio of production and respiration (P/R) provides a comparison of the relative amount of energy channelled into production or maintenance in an ecosystem. The P/R ratio tends to decrease when the ecosystem is under stress, since less energy will be channelled into production and more will be required for maintenance. Organic enrichment, on the other hand, will tend to increase P/R in aquatic systems. In natural communities and ecosystems, it has been shown that toxicant stress will reduce not only diversity, but also production of remaining species, since much of the energy of the surviving individuals is diverted into repair, replacement and adaptation so that less energy can be spared for interspecific relationships necessary for maintaining diversity.

The supply of nutrients is often a limiting factor for primary production (and in turn, for secondary production). Nutrient recycling is undoubtedly one of the most important ecosystem functions. Studies on terrestrial systems and freshwater streams show that stress will impair the ability of the community to retain essential nutrients, and nutrients will be leaked out (exported out) of the system instead of being recycled within.

In a study on stresses under eight toxicants on 25 structural and functional endpoints in microcosms, species composition and species richness were found to be the most consistent structural endpoints, while net daily metabolism (net changes in the dissolved oxygen content of the water over a light/dark cycle) was the most consistent functional endpoint.

Decomposition is an important ecosystem function. Decomposition may be estimated by measuring a variety of enzymes (e.g. amylase, cellulase, glucosidase) relating to decomposition. Metals from smelters have been shown to decrease the decomposition rate of leaf litter. Likewise, coal ash discharge has been shown to decrease leaf decomposition rate in streams. Addition of 5–10 µg L^{-1} cadmium to an artificial stream depresses production, respiration and leaf decomposition, and 40 µmol L^{-1} arsenic decreases decomposition of organic matter in an experimental lake. It should be noted that the carrying capacity of an ecosystem in relation to biodegradable waste depends on decomposition rate, and any impairment or decrease in decomposition will reduce the carrying capacity of the system.

As in subindividual biological organizations, there are also homeostasis mechanisms to enable communities and ecosystems to adapt to toxicant stress and restore their normal status after perturbation. *Resilience* is the innate capacity of an ecosystem to recover from natural or anthropogenic stress, and is defined as the inverse length of time required for the ecosystem to return to the preimpact level. Theoretically, like lower levels of biological organization, populations, communities and ecosystems should also have homeostasis mechanisms to enable them to compensate for toxic effect and restore their original state. Significant ecological changes will not occur if levels of toxicants are below the population tolerance threshold. From an ecotoxicological and environmental management point of view, it is of considerable importance to determine the resilience of ecosystems since knowledge of the resilience of an ecosystem enables us to assess the ecological risk involved. For example an instance of ecological damage is deemed to be acceptable if the damaged system may be able to recover from toxic effects within a few weeks, while the same ecological damage would be unacceptable if the system takes a few years to recover from the same impact. The algicide atrazine reduces phytoplankton abundance and photosynthetic rate, but the phytoplankton community can recover within a relatively short time. After a major oil spillage in Alaska, recovery of the fish population took a year. Likewise, the recovery of benthic communities has been documented after the closing down of pulp mills in Sweden. The time for recovery, however, varies considerably between different populations. Following an oil spillage, recovery of the phytoplankton community in the water column may take only a few weeks, while recovery of an affected seabird population may take decades. In an aquatic habitat, resilience may depend on:

- the existence of nearby resources;
- the transportability of species;
- the condition of physical habitats after perturbation; and
- the persistence of residual toxicants in the environment.

Functionally important species and redundancy

As there are key species that control the structure of community and ecosystems, so certain species play a much more important functional role than others in the community and ecosystem. In the rock intertidal habitat, for example, barnacles and mussels are responsible for over 80% of energy channelling into the system, and therefore they are much more important than the other species in controlling energy flow. Microorganisms play a key role in nutrient recycling in both terrestrial and aquatic ecosystems. Any inhibition of microbial activities is likely to lead to disturbance of the nutrient dynamics of the system and hence serious ecological consequences.

It is generally believed that there is often a considerable amount of functional redundancy in ecosystems. That is, there is more than one species that can perform the same functional role within the same system. Because of this functional redundancy, important functional processes (e.g. energy flow and nutrient recycling) may not be significantly affected by the removal of a few species, since there will be other species available to fulfil or substitute for the same functional role. Thus it is argued that protecting the structural properties of ecological systems is more important than protecting their functional properties. Indeed, there is good evidence to show that ecosystem functions such as primary production, respiration and nutrient cycling are usually less vulnerable to toxicant stress, due to the high functional redundancy and short generation time in the biotic communities that carry out these key functions. The dose of toxicant must be high enough for all resistant species performing the same ecological function to be eliminated before dysfunction of ecosystem will be observable. On the other hand, it is those ecological components with low redundancy and a short life cycle that are particularly sensitive to perturbation. However, once essential ecosystem functions have been impaired, major ecological consequences will occur.

Use of microcosm and mesocosm in ecotoxicological studies

In the natural environment, indirect effects (e.g. species interactions, ecological relationships) are often complex and important, and ecological processes may also be affected by toxicants. Toxicity data derived from a single species therefore, have clear limitations in predicting ecological consequences. On the other hand, the scale of study, slow response, large variability and other inherent difficulties make it difficult to study the direct and indirect effects of toxicants on natural communities and ecosystems. To overcome these problems, one possible approach is to build artificial ecosystems, and use these as tools for studying community and ecosystem responses. These small artificial ecosystems contain an assemblage of living organisms within a defined physical boundary and, depending on their size, may be classified as 'microcosm' (small-size ecosystem) or 'mesocosm' (medium-size ecosystem). Microcosms and mesocosms serve as compact subsets of natural ecosystems and possess the following properties:

• biological components originated from natural ecosystems;
• physically enclosed and isolated from natural ecosystems;
• closed or partially closed to exchange of species and materials with the natural ecosystem;
• important ecological processes such as photosynthesis, respiration, production, metabolism, nutrient recycling, competition, predation, decomposition, etc.; and
• self-regulating homeostasis mechanisms.

Microcosms

In simple terms, microcosms are miniature ecosystems artificially assembled to simulate the processes and interactions of a sector of an ecosystem. The unit normally consists of animals and plants representing different trophic levels. A summary of some of the organisms used in standardized aquatic microcosms is shown in Table 5.2. These systems may range from fairly simple miniature ecosystems consisting of only three trophic levels (primary producers, consumers and decomposers) and a few species in sealed jars or aquaria, to more complex ecosystems consisting of more than a hundred species in outdoor artificial streams, artificial ponds or terraria. In some larger-scale studies, an entire lake may be isolated for dosing experiments. The input to the microcosms or mesocosms may include light energy, mixing energy, water, nutrients, regular addition of organisms, or a combination of the above. Microcosms can be relatively easy to establish, and sustainable for extended periods. Because of their simplicity, replication is also possible. However, due to the self-organization of the system, each microcosm may develop its own unique properties and there

Table 5.2 Some organisms used in the standardized aquatic microcosms.

Algae
Anabaena cylindrica
Ankistrodesmus sp.
Chlamydomonas reinhardi 90
Chlorella vulgaris
Lyngbya sp.
Nitzschia kutzigiana (Diatom 216)
Scenedesmus obliquus

Animals
Daphnia magna
Hyadella azteca (amphipod)
Cypridopsis sp. or *Cyprinotus* sp. (ostracod)
Hypotrichs (protozoa)

can be considerable variability between replicate microcosms, even though all of them were established with the same composition and kept under the same conditions. The ecological processes occurring in the microcosms simulate those in natural ecosystems, although they are simplified because the microcosms are enclosed and isolated. There is also a self-regulating homeostasis in production and consumption within the system. Compared with single-species tests, microcosm tests are closer to reality, simulate better (and hence are a better predictor of) natural changes in species dominance, diversity and trophic relationships, and help to identify sensitive or resistant species in the ecosystem.

While microbial and aquatic microcosms and mesocosms have been developed and used in many situations in ecotoxicological assessment, the use of terrestrial microcosms and mesocosms is virtually nonexistent. The size of microcosms may range from 1 L to 1000 L. The simple system developed by Taube may serve as an example to illustrate a typical laboratory microcosm (see Fig. 5.2). The system consists of: (a) primary producers comprised of some 10 species of single-cell and filamentous algae; and (b) consumers comprised of five species of grazers (*Daphnia*, amphipod, ostracod, protozoa and rotifer) in a 4-L glass container. Chitin cellulose, distilled water, salts and vitamins, and sand are added to the system. This microcosm has been used to test the effects of pesticides. Administration of 10 µg L^{-1} of the pesticide malathion eliminated *Daphnia* after a week. The decrease in the *Daphnia* population reduced the grazing pressure and caused an algal bloom. The system restored itself, however, when the malathion degraded and *Daphnia* was re-introduced.

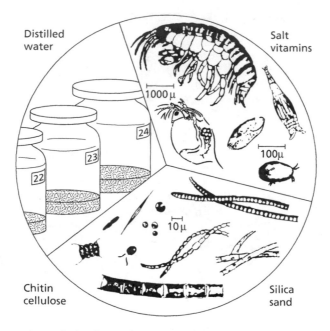

Fig. 5.2 Components of a standardized aquatic microcosm.

Other more complicated microcosms may contain sediment, water, producers and both primary and secondary consumers. A microcosm of this kind was developed by the Oak Ridge National Laboratory of the US, and used to determine safe levels of chemical exposure in a shallow water community. The microcosm consisted of 80 L of water and 10 L of natural sediment, and primarily consisted of a littoral macrophyte community with more than 50 species of algae and macroinvertebrates. The microcosms were dosed with a continuous flow of toxicant, and a number of endpoints including net primary production, P/R ratio, water chemistry and changes in species composition (brought about by interactions amongst sensitive and tolerant species) were measured and compared with those of a control. Persistent ecological changes in both structure and function of the littoral macrophyte community were observed when these microcosms were exposed to toxicants. These included a reduction in P/R ratio (as indicated by changes in dissolved oxygen levels and pH), destruction of the macrophyte community structure and subsequent elimination of dependent species. Data and safe levels derived from this laboratory microcosm system also agreed closely with results derived from a similar experiment carried out in much larger outdoor artificial ponds (mesocosms).

The major advantage of microcosm studies over single-species tests is that both direct and indirect effects on ecosystem structure and function can be determined. Toxic effects on parameters such as primary productivity, species dominance, nutrient recycling, predator–prey relationships and system recovery (e.g. through trophic relationships and competition) can be measured. The results of microcosm studies also allow us to verify results from single-species studies and detect significant indirect effects at system levels which cannot possibly be assessed by means of single-species testing. Another advantage is that the environmental fate of contaminants in the system can be determined, in order to provide valuable information on partitioning of pollutants in sediment, water and the various biocompartments within the system. Information on bioavailability and transfer of contaminants along food chains (and the subsequent ecological effects) is useful in risk assessment. For example, laboratory stream microcosms have been used to demonstrate uptake and bioaccumulation of dieldrin in benthic algae, and the results of the studies showed that dieldrin uptake was directly proportional to current flow.

Good correlation has been demonstrated between the parameters studied (i.e. abundance, species dominance, nutrient dynamics and environmental fate of toxicants) in outdoor ponds and in field studies. In 1987, the US Environment Protection Agency (USEPA) developed a set of guidelines for the use of microcosms. In spite of the fact that many microcosms have been developed since the 1970s, no standardized microcosm has been established for regulatory purposes thus far. This can be attributed to a number of limitations and unresolved technical problems involved in microcosm studies. The major criticism of many microcosm systems is their inability to simulate certain important ecological processes such as larval dispersal and changes in physical dynamics. Also, the high variability in system response makes detection of toxic effects difficult. Some experts have concluded that microcosms appear to represent the worst of both worlds: they are too complicated to give results that can be easily interpreted, but too artificial to be of immediate relevance to field situation.

Mesocosms

The term mesocosm is often taken to mean middle-sized world falling between laboratory microcosms and the large, complex real-world macrocosms. Mesocosms are typically man-made ecosystem segments that can be in the form of outdoor or indoor artificial streams, earthen ponds or enclosures. The size of mesocosms normally ranges between 1 and 300 m^3 in volume, or from 0.01 to 0.1 ha. Sometimes natural ponds or lakes have been used. Because their size is much larger than microcosms, mesocosms typically contain several trophic levels, and an inclusion of higher trophic level predators such as fish and lobsters is also possible. Compared with microcosms, mesocosms are ecologically much more complicated, and therefore provide a much more realistic experimental simulation of natural ecosystems for ecotoxicological assessment. Physicochemical endpoints commonly measured in mesocosm studies may include pH, dissolved oxygen, total nitrogen, total phosphorus, dissolved organic carbon (DOC) and particulate organic carbon; while biological endpoints may include species composition, species diversity, species richness, biomass and productivity of phytoplankton, zooplankton and benthos, ecosystem metabolism, nutrient recycling rate, decomposition, and survival, growth and reproduction of a particular species within the system. The environmental fate of toxicants in various physical and biological compartments and biomagnification may also be studied.

Mesocosms which are constructed as artificial streams may be open, flow-through or closed (recirculating) types, and may contain single species, multiple species at lower trophic levels, or multiple species at several trophic levels. In some studies, outdoor experimental stream channels have been developed into mesocosms to evaluate water quality standards.

A variety of mesocosms with different designs have been developed and employed to study the effects of pesticides for regulatory purposes. Overall, the results of mesocosm studies have shown that outdoor microcosms do respond similarly to larger mesocosm systems when exposed to toxicants. On the other hand, mesocosm studies have clearly demonstrated deficiencies in comparison with microcosm and single-species studies. For example, it has been shown that pyrethroids are extremely toxic to fish and insects in laboratory tests, but because of their hydrophobicity, these pesticides are thoroughly sorbed into sediments and less available when tested in mesocosm studies in the field, although adverse effects on sensitive invertebrates were still found.

There are examples of practical ways in which mesocosms are being used by regulatory authorities to make management decisions. In Germany, artificial streams have been used to study the effects of sewage, nutrients and detergents on stream ecosystems. In the US, mesocosms constructed from 10 000 to 20 000 L tanks have been used to evaluate the impact of pesticides on aquatic communities. Cylindrical enclosures made of polyethylene, PVC or rubber suspended in the water column from a flowing platform have been constructed to study the effects of a variety of toxicants (e.g. oil, nutrients and metals) on plankton dynamics, benthic community processes and fish growth. An international consortium of institutions has conducted experiments using the Controlled Ecosystem Pollution Experiment (CEPEX) which consists of 1335 m^2 enclosures in which natural seawater and plankton have been enclosed. In this system, the long-term effects of sublethal stress of pollution on marine planktonic ecosystems have been studied.

Problems in using microcosms and mesocosms in ecotoxicological studies

Although there are major advantages in microcosm and mesocosm studies, there are also clear limitations and inherent problems in this approach. These limitations and problems have attracted considerable criticisms from ecotoxicologists.

First, there are major limitations in successfully simulating the temporal scale, the spatial scale and the ecological complexity in mesocosms. Most of the microcosms and mesocosms developed are too small, and their ability to simulate complex ecological relationships and reflect environmental realism is questionable. Small mesocosms (< 10 m³) are well known for their inherent instability, and rapid natural changes in community structure may be expected. In one mesocosm study it was shown that biological variations in small (2 m³) ponds were so great that it was impossible to differentiate the effects of toxic waste water discharge from natural variations. Also, the small size of microcosms and mesocosms cannot support extensive and repeated destructive samplings. However both microcosms and mesocosms are often unstable over long periods of time, while many indirect effects and relationships are only detectable after prolonged periods. While microcosms and mesocosms can partly simulate ecological relationships, they may fail to simulate certain important relationships and therefore may lead to erroneous conclusions. For example, the container (delineation) of the mesocosm can reduce wind and wave action; and difficulties in simulating air/water and physical mixing processes (e.g. gas exchange, vertical mixing, current) are well recognized. Also, the biological communities within the mesocosms and microcosms differ from those in the surrounding natural environment. The exclusion of nutrient input and exchange and larger organisms has been shown to lead to quite different experimental results. Nutrient depletion inside the container and the growth of periphytons on the walls of the enclosure are also phenomena that do not occur in the natural environment. Proper scaling of animals and plants, especially those at higher trophic levels is difficult. It would be impractical to include predators of a higher trophic level (e.g. predatory fish) in the enclosure, because the productivity from the relatively small volume of the mesocosm would not be able to support fish growth. So far, manipulation of higher trophic levels has been unsuccessful in mesocosm studies. For example, it would be very difficult to include predators such as rainbow trout even in a 10-m diameter mesocosm, since the typical density of the animal is around 1 in 1000 m² in the natural environment. While most mesocosm studies have been 'one treatment, one mesocosm', there are typically large variations between mesocosms after a short while, despite the fact that they can be constructed in an identical way at the beginning of an experiment. For example, in the mesocosm developed by Monticello, 85% of all macroinvertebrates were isopods in one stream, 67% were chironomids in another and 69% were amphipods in a third. The practical difficulties of having a large number of replicates to cover the variance limits the use of mesocosms in ecotoxicological studies. On the other hand, the limited replicate numbers do not allow us to extrapolate the mesocosm results to natural systems with great confidence.

Field evaluation

The ultimate goal of ecotoxicological studies should be to predict the ecological consequences in order to provide risk assessment at the community and ecosystem levels. The exposure regime in laboratory tests is normally at a continuously high concentration, which does not reflect a realistic environmental exposure, since levels of toxicant are seldom uniform in the natural environment. Furthermore, many other factors in the natural environment may also affect the availability and toxicity of toxicants. For example, suspended solids and organic carbon in natural water may greatly reduce the bioavailability of lipophilic compounds, and photolysis and microbial degradation may also reduce the concentration of toxicants. To cover these uncertainties, 'safety factors', or application factors, are usually applied in deriving the 'safe concentration' of a chemical in the field. It must be stressed that laboratory results at organismal levels and below are relevant only if they can be extrapolated to make predictions on a population, community or ecosystem, or can be used to validate field observations. Obviously, results of single-species tests cannot provide any information on important community and ecosystem responses such as diversity, stability, resilience, nutrient dynamics, trophic relationships and species interactions that may have a significant impact on complex multi-species field conditions. It is generally held that simple laboratory tests tend to overestimate the magnitude and duration of exposure to a toxicant relative to that observed under natural field conditions. It has been concluded that the effects of nonpersistent, acutely toxic chemicals in field tests can be accurately predicted from laboratory studies. However, if there are rapid changes in the exposure concentration of a toxicant, a single-species test may overestimate the potential effects in the field. This is because animals in the laboratory are directly exposed to unrealistically high concentrations of toxicants within a short period of time, and also because toxicants are subject to photooxidation and their bioavailability and persistence are likely to be reduced under field conditions. On the contrary, laboratory tests tend to underestimate the potential impact of chronic toxicants that require metabolic activation or involve biomagnification. There are clear exceptions; for example, some toxicants increase their toxicity under field conditions: anthracene is not acutely toxic to animals and plants at low concentrations, but the toxicity of this PAH is increased by some 50 000 times after exposure to ultraviolet through normal sunlight. Once again, this example serves to illustrate the importance of conducting field tests to validate laboratory results.

Despite the obvious importance of field validation, relatively few well-designed field experiments have been carried out to validate laboratory predictions and extrapolations. There are two possible approaches in field validation. The first approach involves field monitoring at polluted sites. This approach suffers the limitation that the potential interactions of a large number of other factors are not known and cannot be controlled. The second approach involves field experimentation on the whole ecosystem. Relatively few studies have been carried out using this approach to validate toxicological models. Few field experiments have been carried out on whole natural ecosystems, because of the large expense involved, difficulties in replication and the long study period required. Also, scientific manipulation of whole ecosystems is only possible in a few locations; for example, in isolated areas relatively free from human disturbance.

Conclusions

Much ecotoxicological research and its applications has focused on single species. The major disadvantage of the single-species approach is its inadequacy in predicting the effects and hazards of a chemical on natural systems, since natural systems are far more complex than any laboratory testing system. Ecological factors and their interactions such as trophic relationships, species interactions, nutrient dynamics, limiting factors, environmental fate of chemicals in the natural systems, etc. are poorly understood. Furthermore, a great variety of physical, chemical and biological factors in natural conditions are difficult, or impossible, to simulate under laboratory conditions. Yet many of these factors may significantly alter the form, bioavailability, quantity and toxicity of toxicants in the natural environment. These uncertainties pose questions as to the validity of extrapolation or prediction-making from laboratory test results.

Most of the early ecotoxicological studies focused on the responses of suborganisms. Relatively few studies have been carried out on and little is known about population, community and ecosystem responses, and current predictive power at the population level and above is relatively weak.

Microcosms and mesocosms are likely to provide more appropriate measures for the establishment of ecologically relevant parameters in relation to no-effect concentrations of a toxicant. Major shortcomings of these experimental systems include lack of replication, high cost and difficulty in manipulation and control of the experimental conditions. Nevertheless, these studies offer useful information to help to identify and focus on important parameters and processes to be monitored and studied. They also provide a valuable means of calibration and validation of results derived from bioassays. These systems also help to explain and validate experimental results derived from lower levels of biological organization.

Further reading

Beyers, R.J. & Odum, H.T. (1993) *Ecological Microcosms*. Springer-Verlag, New York.
Boudou, A. & Ribeyre, F. (eds) (1989) *Aquatic Ecotoxicology: Fundamental Concepts and Methodologies*, I and II. CRC Press, Boca Raton.
Cairns, J. Jr (1986) *Community Toxicity Testing*. American Society for Testing and Materials, Ann Arbor, Boca Raton.

Calow, P. (1996) Ecology in ecotoxicology: some possible 'rules of thumbs'. In: *Ecotoxicology: Ecological Dimensions* (eds D.J. Baird, L. Maltby, P.W. Greig-Smith & P.E.T. Douben), pp. 5–11. Chapman & Hall, London.

EPA (1987) Toxic substances control act test guideline. *Subpart-D-Microcosm Guidelines. Federal Register (USA)* **52** (187), 36363–36371.

Giddings, J.M. (1986) A microcosm procedure for determining safe levels of chemical exposure in shallow communities. In: *Community Toxicity Testing* (ed. J. Cairns). ASTM publication 04-920000-16, Philadelphia.

Giesy, J.P. & Allred, P.M. (1985) Replicability of aquatic multispecies test system. In: *Multispecies Toxicity Testing* (ed. J. Cairns). Pergamon Press, Elmsford, NY.

Rand (1995) *Fundamentals of Aquatic Toxicology. Effects, Environmental Fate and Risk Assessment.* Taylor & Francis, Washington, D.C.

Taub, F. (1989) In: *Aquatic Ecotoxicology: Fundamental Concepts and Methodologies* (eds A. Boudou & F. Ribeyre), pp. 47–92. CRC Press, Boca Raton.

6: Dose and Concentration Response Relationships

Introduction

A toxicant is an agent that can produce a significant adverse response (effect) in a biological system, causing damage to its structure and function or, in extreme cases, death. The impact of a chemical on an ecosystem is initiated when the chemical enters the environment. The fate of a chemical following its entry depends on a number of factors, including:

1 the physicochemical properties of the substance;
2 the physical, chemical and biological properties of the ecosystem; and
3 the origin, amount, chemical form and fate of the substance in the environment.

The proportion of the total amount of a chemical present in the environment that can take part in physicochemical and biological processes is referred to as its environmental availability. In most cases, only a portion of this environmentally available fraction eventually enters the biological system. This is usually achieved through either passive diffusion or active transport across biological membranes. The proportion of the total quantity of an environmentally available compound that is available for biological action or involved in metabolic processes in a biological system is defined as the bioavailability of the chemical.

Due to different physicochemical properties various chemicals may have different impacts on separate components of a biological system. Once inside an organism, toxic chemicals may be distributed by the circulatory systems to various parts of the target organism. Some chemicals, such as strong acids and alkalis, exert their toxic effect in a nonspecific way by denaturing proteins and dissolving tissues. More commonly, however, toxic chemicals affect specific components of the biological system. Indeed, a chemical can only exert its toxic effect when it is bound to a specific site of toxic action (receptor). Receptors may be parts of cells confined to certain tissues, or can be nucleic acids or specific regions of proteins within nerve synapses or membranes distributed among the cells, of an organism. For example, contraction of certain muscles is mediated through a neurotransmitter, acetylcholine, and muscle contraction is initiated when acetylcholine binds to a receptor on the muscle cells. In a normal sequence of events, subsequent degradation of acetylcholine by an enzyme, acetylcholinesterase, returns the muscle cells to their resting state. Organophosphate pesticides and certain neurotoxins can disrupt the normal signalling between neurones and muscles either by binding directly to the acetylcholine receptors on the muscle cells or by blocking the action of acetylcholinesterase. The resultant uncontrolled muscular contraction causes muscle fatigue and paralysis and, in extreme cases, fatality.

Dose–response relationships

In establishing a true dose–response relationship, it is necessary to distinguish between the concentration of a chemical in the environment and the actual amount that reaches a target tissue. However, as it is usually very difficult to determine the actual amount of a chemical reaching a target tissue, the total amount that enters the body is often used instead. In general, the biological effect of a toxicant is assumed to be related to the dose administered. The assumptions are:

1 that there is a particular molecular site (receptor site) with which the chemical reacts;

2 that the response is proportional to the concentration of the chemical bound to the site of action; and

3 that the concentration of the compound is related to the dose administered. It is important to note that in examining dose–response relationships, the response (effect) must be readily quantified in a reproducible way that is relevant to the toxic processes under investigation. As the impact of a toxicant on an organism is likely to be related to the size of the organism, dose needs to be measured in terms of concentrations rather than absolute amounts. In animal experiments, doses are often expressed in weight or molecular units per kilogram of body weight (i.e. mg/kg body weight or mg/kg/body weight/day) or per square metre of body surface area. It is important to distinguish between several terms that are commonly used to represent concentrations of chemicals affecting an organism. Exposure is the concentration of material in the air, water or soil to which an animal is exposed (note that this is different from the dose received by an organism). Dosage is a term incorporating dose, frequency and duration of dosing. A critical factor in toxicity determination is the exposure duration. The toxic effect of even a low toxic dose can be significantly increased by extending the exposure duration.

Conceptually, low toxicant concentrations may produce no observable effect, but as the concentrations increase beyond a critical level, an increasing adverse effect can be observed, finally reaching death (Fig. 6.1). A distinction can be made between essential substances which become toxic at higher levels, such as many trace metals, and substances which are toxic at all levels, such as DDT. To assess the toxicity of a compound on a biological system, an observable and well-defined effect (endpoint) must be identified. Individual organisms may respond differently even to the same stress as a result of genetic or phenotypic variations. Consequently, an evaluation of the toxic potency of a compound will require the use of statistical techniques.

In the past, assessment of toxicity involved an estimation of an 'all-or-none' response, often measured as lethality, of a chemical with organisms.

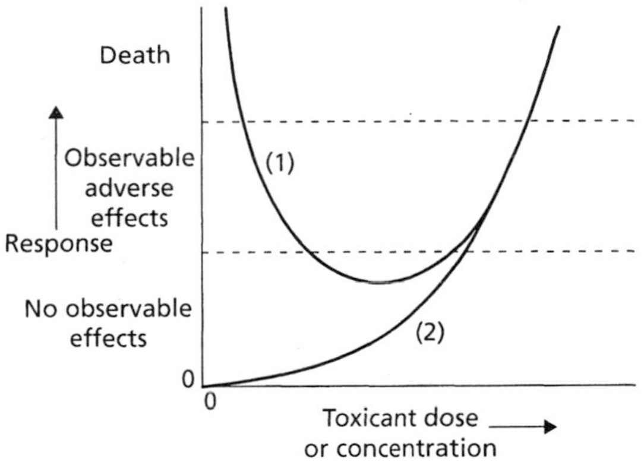

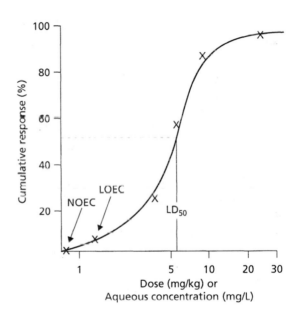

Fig. 6.1 Biological responses to differing doses or concentrations of (1) essential substances which become toxic at higher than essential levels and (2) nonessential toxic substances.

Fig. 6.2 Cumulative dose–response curve. In a lethality experiment the response is the cumulative percentage of animal mortalities with the actual data points indicated as crosses. Lowest observable effect concentration (LOEC) and no observable effect concentration (NOEC) are indicated.

Mortality measurements are used as they are easily definable, biologically interpretable and statistically repeatable. They provide a convenient means of comparing the toxicity of a wide range of substances without requiring a prior understanding of the mechanism involved. Tests that are based on mortality, or survival, are used primarily to examine the short-term (acute) effects of a toxic exposure. When considering toxicity to aquatic organisms, initial experiments are generally performed to obtain an indication of the concentration range in water that should be studied in depth. Following these range-finding tests, several groups of organisms (usually 10 or multiples of 10 per group) are exposed to an increasing, narrower range of concentrations of the chemical. In these so-called 'definitive tests', the number of dead organisms in each group is recorded after a fixed period of exposure, e.g. 96 h. The cumulative percentage of mortality at each concentration is estimated and plotted as in Fig. 6.2.

Percentage	Probit
10	3.72
20	4.16
30	4.48
40	4.75
50	5
60	5.25
70	5.52
80	5.84
90	6.28

Table 6.1 Conversion of percentage into probit units.

The graph is typically sigmoidal, with the centre portion of the plot nearly linear (i.e. the effect of the chemical in this segment is almost directly proportional to the concentration), while the upper and lower ends of the curve asymptotically approach 100% and 0%, respectively. Further analysis of the curve in Fig. 6.2 reveals that the data points in the central segment are least affected by variations in the observed percentage mortality, while points in the asymptotic regions are most sensitive to a small deviation in the observed mortality value, resulting in a large error in the estimation of the corresponding concentration or dose. On this basis, the toxic potency of the chemical is commonly expressed or summarized using one of the following terms, depending on the nature of the experiment.

1 LC_{50} is defined as the concentration at which there is 50% mortality of the organisms.

2 LD_{50} is defined as the dose at which there is 50% mortality of the organisms.

3 IC_{50} is defined as the concentration at which 50% of the growth or activity is inhibited.

4 EC_{50} is defined as the concentration at which 50% of the predicted effect is observed.

The EC_{50} and IC_{50} should be lower than LC_{50} as impairment of function should occur before death. Based on the above principle, LC_{25} and LC_{75}, corresponding respectively to 25% and 75% mortalities, can also be estimated. The expression chosen would depend on the type of information required. Although the LC_{50} or LD_{50} can be read from the sigmoid graph, it is more useful if the median lethal concentration or dose can be estimated mathematically. One way of achieving this is by probit transformation which converts the percent response figures into probit values (see Table 6.1). Using this data the sigmoidal dose–response curve can be converted into a straight line as shown in Fig. 6.3 thus allowing statistical treatment by the linear regression techniques. It should be noted that not all toxicity data can be successfully described by a log-probit model, and a chi square goodness-of-fit test can be used to ascertain whether such a model is appropriate. In some cases, transformation other than a probit, such as logit or angle, may give a better fit.

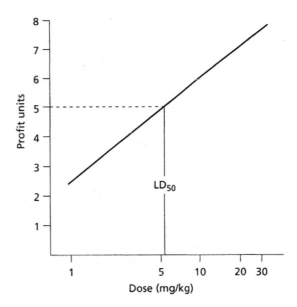

Fig. 6.3 Probit transformation of a dose–response curve.

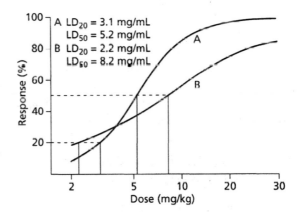

Fig. 6.4 Comparison of dose–response curves with different slopes for compounds A and B.

Toxicity of a compound, expressed as LC_{50} or LD_{50}, is useful for comparative purposes; the smaller the LC_{50}, the more toxic the compound. Acute toxicity data are often compared by evaluating the magnitude of difference in LC_{50} values. However, the dose–response curves may have different shapes and different slopes from the central linear section. This leads to variations in relative toxicities of the compounds as shown in Fig. 6.4 with the dose or concentration.

These tests designed to evaluate the relative toxicity of a chemical for selected organisms upon short-term exposure (e.g. 48 h or 96 h) to various concentrations of a test chemical are referred to as *acute tests*. Common endpoints are mortality, immobility or inhibition of growth. Despite the popularity of mortality-based acute tests in ecotoxicological assessments, there is a growing interest in investigations concerned with the sublethal effects of environmental toxicants. A chemical that does not cause the death of test organisms within the test duration may still have a long-term deleterious impact on the test organisms, e.g. through an impairment of future reproductive output, or reduced growth rate or postexposure survival. An assessment of these chronic sublethal effects is important in providing an early warning to prevent irreversible damage to biological systems and in ecological risk assessment exercises. Indeed, continuous exposure to a low toxicant dose over a long period of time can result in chronic toxicity. In a *complete chronic toxicity test*, the test organisms are exposed to relatively low toxicant concentrations (at least five different concentrations should be used) for an entire reproductive life cycle (e.g. egg to egg). Test methods for chronic toxicity tests are often highly variable and complicated as they must take into account the life history of the individual test organism. To reduce the cost and time needed for these tests, full life-cycle tests are often replaced by partial life-cycle tests, which involve only several sensitive life stages (e.g. embryo to larva). In order to formulate meaningful tests for assessing chronic toxicity, important biological information about the test organisms, such as life cycles, nutritional and physiological requirements, behaviour, predator–prey interactions, physical requirements (light, current, photoperiod, substrate, etc.), needs to be collected. Chronic toxicity arising from exposure to sublethal concentrations of toxicants is often integrated and reflected in key life-history parameters (e.g. growth rates and reproductive efforts), or patterns of energy allocation of the organism. Consequently, it is assumed that an assessment of such effects will provide important information on the long-term influences of the test chemical on individual organisms as well as possible ecological impacts on natural populations and communities.

Toxicological effects of continuous and intermittent exposures

Environmental pollution can be attributed to many sources, such as accidental spillage of chemical wastes, periodic precipitation contaminated with airborne pollutants and the discharge of industrial or sewerage effluents. Clearly, contaminants entering the environment via these sources tend to be variable in their volume and chemical nature. For example, the quantity and chemical composition of industrial waste waters are closely linked to the production cycle of the industrial processes producing them and, due to environmental conditions, organisms in the receiving environments will rarely be exposed to continuous or consistent doses of toxicants. Furthermore, the environmental impact of these discharges can be significantly modified by the mode and degree of mixing in the receiving waters as well as dilution by rainfall.

Most pollution events are episodic and discontinuous, but toxicants used in rigidly controlled laboratory tests are often held at fixed concentrations, rarely representative of field exposure levels. Consequently, it is difficult to extrapolate data obtained from laboratory-based toxicity tests to actual ecosystems. Indeed, such an approach can only be of limited value if the ultimate aim is to predict realistically the biological effect of the toxicants or pollutants on an ecosystem. A meaningful assessment of the toxic effects of an intermittently discharged pollutant depends ultimately on the ability to predict its effect on the receiving environment. One approach to determining the impact of intermittent pollution events is to estimate toxicant burden in an organism and correlate toxic responses (acute or chronic) with tissue toxicant accumulation.

There is growing evidence that mortality, for example, is correlated with whole-body residues of pollutants in at least some organisms. For example, the survival of sticklebacks (*Gasterosteus aculeatus*) and mummichogs (*Fundulus heteroclitus*) is inversely related to cadmium (Cd) levels, and zinc (Zn) and copper (Cu) residues in the body tissues. It is possible to conceive from these results that an estimation of the pollutant concentration in different tissues of the target organisms may be a useful predictor of the biological impact of the pollutant on the organism, assuming that the underlying mechanisms of accumulation are not different in fluctuating as compared to continuous exposures.

Additive, synergistic and antagonistic interactions of different toxicant species in a mixture

Poison accumulation assays assume that a toxicological response is elicited by the accumulation of the test substance above a threshold level in the tissues. One potential problem with this approach is in the case of complex effluents which may require identification of all the pollutants in the target tissue or organ. Indeed, toxic chemicals rarely occur in isolation in natural environments. Also, exposure to mixtures of chemicals may result in toxicological interactions. A *toxicological interaction* refers to the phenomenon whereby exposure to two or more chemicals results in a biological response quantitatively and qualitatively different from that expected from the action of the individual chemicals alone. Examples of toxicological interaction include synergism and antagonism. *Synergistic interaction* refers to the situation where the combined effect of two chemicals is greater than the sum of their individual effects, while *antagonism* refers to interference between two chemicals when applied together which results in a reduced impact. Interactions may also occur between a toxicant (e.g. a metal), and an environmental parameter (e.g. pH). Two mechanisms have been proposed for the influence of pH on metal toxicity: mobilization and competition. Toxicity of a metal is directly related to its solubility or the bioavailability of free metal ions in solution. Increased acidity can mobilize metal ions from a bound sequestered state to an active free ion state, resulting in increased toxicity. In the competition mechanism, hydrogen ions may displace metal ions from their binding sites on the cell wall as pH is lowered. The mobilization mechanism implies that metal toxicity should increase with lowered pH, while the competition mechanism suggests that it should decrease with lowered pH. Further complications may arise as toxicological interactions may potentially be dose related. On this basis, the predicted impact of chemical mixtures based on individual effects of each toxic species is likely to be erroneous. Clearly, an understanding of the synergistic and antagonistic interactions between individual toxic species at different concentrations is important in assessing the toxicity of complex chemical mixtures.

Factors affecting toxicity

In general, water-soluble (hydrophilic) chemicals are more readily available to organisms than water-insoluble (hydrophobic) chemicals. Water-soluble chemicals may enter an organism through the general body and respiratory surfaces. Chemicals in food are usually taken up through the gastrointestinal wall following oral ingestion. Indeed, there are three main mechanisms by which pollutants are transported into cells: passive transport or diffusion, filtration and active transport. Cell membranes are composed of two protein backbones separated by a phospholipid bilayer. The membrane structure is interrupted by invagination of the protein molecules to form pores through which chemicals can penetrate. Neutral molecules, particularly lipophilic substances, tend to be transported passively through biological membranes. Absorption of charged molecules, e.g. acids and alkalis, depends on their degree of ionization which, in turn, depends on the pH of the surrounding medium. The process can be mediated through either filtration, which depends on the size of the molecules, or active transport, which requires energy. Active transport is usually used to move hydrophilic molecules, via a carrier molecule, against an electrochemical gradient across a biological membrane. In this case, the toxicity of a chemical compound is dependent on its physicochemical properties.

Toxicants taken up by living organisms from their surroundings or from food sources are either metabolized to more soluble forms and excreted, or they are stored in various tissues of the body and prevented from taking part in active metabolism and disrupting vital biological processes in the organisms. It is assumed that toxicants often do not elicit any deleterious effects as long as they are sequestered in a nondividing tissue. The accumulation of toxicants in the body storage zones gradually results in a toxicant concentration in the organism higher than that of the surrounding medium. This involves at least two separate processes:

1 bioconcentration which is the uptake of toxicants by an organism directly from the surrounding medium; and

2 biomagnification which is the uptake of toxicant through the food.

The degree of bioaccumulation of a specific toxicant depends on:

1 rate of uptake;

2 solubility and lipophilicity of the compound;

3 rate of metabolism; and

4 rates of excretion and depuration.

Different species have different tolerance to various chemicals. In the past, the prime criterion for the selection of an organism for toxicity testing was its high degree of sensitivity to chemical compounds. Besides sensitivity to test chemicals, other aspects are important in selecting organisms to be used in a toxicity test. Ideally, these organisms should have the following attributes:

1 reasonably abundant in typical receiving environments;

2 easy to culture in a laboratory setting; and

3 a reliable predictor of the response of other organisms in a community.

When using organisms in toxicity assays, it is important to minimize variations within each treatment group as effects can only be demonstrated by a comparison of variances within and between the groups. However, even when a single species is used, a number of factors such as size, sex, age or developmental (e.g. moult) stage, diet or nutritional status, state of health, reproductive state, etc. can also affect the response of individual organisms to a specific toxicant.

Larvae or juveniles often appear to be more susceptible to chemical stressors than their adult counterparts. This may be due to differences in the degree of development of detoxification systems between juvenile and adult organisms, and/or a higher surface-to-volume ratio in juveniles relative to adults. Furthermore, there may also be age-specific differences in rates of excretion of toxic chemicals. Despite this, embryos of some species are less sensitive to toxicants than adults, possibly due to the extra protection given by a 'semi-impermeable' membrane. It should be noted that dietary factors can affect an organism's tolerance to toxic stress via changes in body composition, physiological and biochemical functions and nutritional status. For example, starved organisms are usually more susceptible to environmental stressors. Presumably, they have less energy and resources at their disposal for dealing with the chemicals either by metabolizing and excreting the substance and/or by evoking more complex detoxification mechanisms. Another complication is the influence of parents on the level of tolerance in their offspring. Very often strong mothers can give rise to stronger offspring which may conceivably be less susceptible to toxic stress. For this reason, field-collected animals are required to be acclimatized under specific sets of environmental conditions, usually in the laboratory, before they can be used in toxicity tests so that any differences in the organisms arising from heterogeneity in the parent environments will be eliminated, or at least minimized. In standard toxicity tests, only healthy and unstressed organisms should be used, as organisms in poor health or pre-exposed to stress may be more sensitive to a particular stress. It is also important to monitor the response of laboratory cultures from time to time to ensure that maternal or environmental effects in the case of newly collected organisms, or the impact of 'laboratory selection' for old cultures are effectively controlled. Indeed, microevolutionary changes in supposedly genetically identical clones of test organisms (e.g. *Daphnia* cultures) have given rise to vastly different test results in separate laboratories.

In addition, environmental parameters, such as temperature, pH, water quality (hardness or softness), salinity, current, substrate and dissolved oxygen can also influence LD_{50} and LC_{50} values. To minimize possible effects of environmental conditions on results of toxicity tests, attempts are made to standardize testing media and conditions as far as possible. For example, in aquatic toxicity tests, synthetic test media prepared by dissolving the appropriate salts in distilled water are commonly used.

Toxicity testing in the laboratory

Toxicity testing in the laboratory usually follows a stepwise tiered approach, progressing from simple short-term tests to more complex and sophisticated long-term tests based on the results of previous tests. Although the details of the protocol for each test may differ, the general test design is similar and can be summarized as follows.

1 Test conditions, such as pH, temperature, dissolved oxygen concentration, photoperiod and chemical composition of test medium, should be carefully controlled.

2 Test organisms are exposed to increasing concentrations of the toxicant in identical test chambers, such as beakers, glass tanks, etc., for a fixed duration.

3 Standardized toxicological endpoints, such as growth, reproduction and mortality, are measured and compared among treatment groups and against the control.

4 Three basic types of controls should be set up.

(a) A negative control in which test organisms are exposed to the same test medium as the treatment groups, but without the test substance or solvent carrier.

(b) A solvent/carrier control may be required in situations where a poorly soluble test substance is first dissolved in a small amount of organic solvent (e.g. acetone, dimethyl sulphoxide—DMSO) in preparing the stock solution. The solvent control is essentially identical to the negative control, but with a solvent/carrier concentration equivalent to the maximum concentration of the solvent present in the treatment solutions.

(c) A positive (reference) control is provided by exposing the test organisms to a toxicant or toxic agent known to produce a well-defined response on the test organisms. The aim is to ascertain the health and sensitivity (hence suitability) of the organisms to be used in the test. On this basis, reference toxicants should be toxic at low levels, stable under testing conditions and relatively nonspecific. Typical reference toxicants include organic (e.g. sodium pentachlorophenate and phenol) and inorganic (e.g. cadmium chloride, copper sulphate) chemicals.

By convention, the test results are only valid if percentage mortality in the control is less than 10%. Control mortality greater than 10% usually indicates possible problems related to handling stress, disease or poor test conditions. In standard toxicity tests, one or a limited number of species is used due to various economical and practical constraints. As the ultimate purpose of these tests is to try to use the test data to predict potential impact of the toxicant(s) on the ecosystem, careful selection of the test organisms is vitally important. A number of criteria/principles should be considered.

1 A group of organisms, preferably at least one from each trophic level (e.g. algae, macroinvertebrates and fish) should be used instead of a single species whenever possible.

2 Test organisms that show observable responses over a wide range of toxicant concentrations are preferred.

3 Test organisms with a wide geographical range and availability are preferred.

4 Whenever possible, toxicity tests should be performed using local or native species.

5 Species that are ecologically important (in terms of taxonomy, trophic level or niche) are preferred.

6 Test organisms that occupy a position within a food chain leading to man or other important species should be selected.

7 Species that have important commercial or recreational value are obvious candidates.

8 Species with life-history characteristics that are amenable to easy maintenance in laboratory cultures, such as short life span, short life-cycle phase, fast growth rate, and resistance to disease, parasitism and physical damage, should be used.

9 Species that can be identified easily and for which there is adequate background biological (e.g. physiology, genetics, ecological niche) and toxicological information should be selected as this will facilitate subsequent data interpretation and utilization.

Although all the above criteria are desirable, it is often difficult, if not impossible, to find a test organism that can fulfil all the requirements. Selection and use of a particular test species would depend on the question asked or the purpose of the experiment, and possible limitations should be made explicit. Apart from selection of suitable test organisms, the exposure systems used can also have an important influence on toxicity results. Basically, four techniques are commonly used.

1 Static tests involve exposing the test organisms in still test medium without any change of medium for the duration of the test.

2 In a recirculation test, test medium is usually pumped into the test chamber from a reservoir, returned and circulated within the closed system.

3 A renewal test is essentially similar to a static test except that the test medium is completely or partially renewed at fixed intervals, say, every 24 h.

4 For flow-through tests, the test solution is passed through the test chambers, usually from a large reservoir, and the medium is not returned after it has passed through the test chamber.

Table 6.2 Advantages and disadvantages of exposure systems.

Exposure systems	Advantages	Disadvantages
Static tests	Technically simple Relatively inexpensive Suitable for acute tests, particularly for screening Lower time and space requirement Minimum problem with waste disposal due to small volume involved	Difficult to ascertain the actual (effective) concentration of toxicant due to: volatility; biological or photolytic degradation; plate out (e.g. metals); influence of confounding variables (e.g. metabolic products and changes in water quality) Less suitable for longer-term subchronic and chronic tests Only suitable for organisms from lentic environments
Renewal tests	Similar to static tests Ensure a toxicant level closer to the nominal concentration than static tests	Similar to static tests May cause undesirable disturbance to test organisms Labour intensive
Recirculation tests	May reduce stress such as low oxygen tension in test medium	Relatively expensive May require a more sophisticated set-up Uncertainty of the effect of the apparatus used in moving the medium
Flow-through tests	More uniform and stable test conditions Results may be more repeatable Suitable for acute and chronic tests Suitable for long-term tests Best for testing chemicals with high oxygen demand Can use organisms from lotic environments	May involve a complex delivery system May involve a high cost due to expensive construction and operation, and the larger amount of material used Problem with waste disposal due to large volume involved

A comparison of the advantages and disadvantages of the four exposure systems is given in Table 6.2.

In all toxicity tests, precise information must be given on the type and concentrations of toxicant used, organisms to be tested, length of exposure, test situations and criteria of effects (endpoints) to be evaluated. Nominal concentrations are often reported which are the concentrations expected using the calculated amounts of test compounds. Regardless of which technique is used, it is always desirable to measure the actual concentrations to which test organisms are exposed by chemical analysis. Nominal concentrations can be misleading at times as effective concentrations of the test chemicals may be reduced through degradation, volatilization, adsorption, etc.

Over the years, a large number of tests have been developed for toxicity tests. National and international organizations (e.g. environmental protection agencies, testing laboratories) that deal with toxicity assessment have recognized the need to standardize the testing protocols in order that:

1 tests selected and performed by different laboratories can be compared for interlaboratory calibration;

2 accuracy and reliability of data can be increased such that data obtained by different laboratories can be shared; and

3 tests can be replicated and conducted by a variety of personnel, and suitable quality assurance procedures implemented accordingly.

Despite the importance and obvious advantages of adopting standardized testing procedures, there are problems in ensuring that 'standardized' protocol actually produce repeatable results. This is evident from a number of interlaboratory calibration exercises (ring tests) undertaken by contract laboratories in the Organization for Economic Cooperation and Development (OECD), which generated unacceptably variable results for 'identical' samples.

To estimate the sublethal effects of toxicants, specific tests are designed to measure the following doses.

1 The maximum dose which does not produce a statistically significant harmful effect. This is expressed as the no observable effect concentration (NOEC) or no observable adverse effect concentration (NOAEC).

2 The minimum dose used in a test which produces a statistically significant deleterious effect, expressed as lowest observable effect concentration (LOEC) or lowest observable adverse effect concentration (LOAEC).

These are illustrated in Fig. 6.2. When levels (L) are used instead of concentrations, NOEC, NOAEC, LOEC and LOAEC are replaced by NOEL, NOAEL, LOEL and LOAEL, respectively. It should be noted that the values of NOEC and LOEC depend on the actual concentrations used in the test. As it is not possible to test an infinite number of intermediate concentrations, the maximum acceptable toxicant concentration (MATC) can only be reported as being greater than the NOEC and less than the LOEC (i.e. NOEC < MATC < LOEC). The MATC is sometimes calculated as the geometric mean of the LOEC and NOEC for regulatory purposes.

Although toxicity tests are useful in providing information on the critical toxicant levels that will result in mortality of the test organisms, there is a need to utilize this information to predict 'safe' concentrations that are applicable to environmental regulations or legislation. One approach is the use of *application factors* (AFs). This involves applying an arbitrary factor to the toxicological data (e.g. by multiplying the ratio between the empirical values of EC_{50} and no-effect concentration by a factor ranging from 0.1 to 0.0001) to obtain a long-term safe concentration for the organisms in the receiving environment. As indicated above, arbitrary factors (e.g. 0.1, 0.01, 0.001, 0.0001) are typically used as the first step when data on long-term or sublethal effects of toxicants are not readily available. The use of arbitrary AFs has a number of shortcomings:

1 LC_{50} values are often not the best estimate for the critical lethal concentrations;

2 the scientific basis for the use of particular AF is often very weak; and

3 such derivation of a 'safe concentration' from data based on lethality presents conceptual problems, and may be invalid.

The mechanisms involved in lethal and sublethal responses may be different. There have been suggestions that chronic toxicity data alone should be used in establishing MATC.

With the availability of more data on the chronic effects of chemicals, AFs can be derived empirically as the ratio of MATC (usually in terms of effects on growth, reproduction, etc.) to the LC_{50} value (usually obtained from a 48-h or 96-h acute test) (i.e. AF = $MATC/LC_{50}$). On this basis, the AF provides a crude estimate of the relationship between the chronic and acute toxicity of a test chemical. The inverse of AF is defined as the acute to chronic toxicity ratio (ACR). In the absence of a suitable data set, the AF value derived from one chemical is sometimes used to predict the AF of another. The assumption here is that AF is relatively constant. However, experimental data have confirmed that AFs can vary by two orders of magnitude between organisms exposed to a single chemical and by four orders of magnitude between different chemicals for a single organism.

Effects on individuals observed in laboratory testing

Mortality

Lethality is an 'all-or-none' or quantal response that is relatively easy to determine, and thus is often used to provide a rapid means for an initial assessment of toxicity. However, it is sometimes necessary to include a postexposure observation period where organisms are placed in a 'clean' environment, and observed for a sufficiently long period of time to ensure that the organisms are indeed dead. This is particularly important if the test chemical has an anaesthetic effect. In toxicity tests, it is common to define death as the failure to respond to external stimulus. Substances that cause the death of a substantial number of individuals in a population will clearly have a direct impact on population size. On this basis, mortality-based toxicity data are relatively easy to interpret and often used for environmental regulation. A large volume of acute toxicity data (LD_{50} or LC_{50}) now exists for a wide range of chemical compounds, and the database is useful for comparing the toxicities of different substances and for risk assessment purposes.

Growth

Growth is an important fitness characteristic of individual organisms, and is easily measured. Growth also tends to integrate and reflect all sublethal effects that may be operating on an animal, and has an overall impact on the success of natural populations. On this basis, growth is routinely measured in most chronic experiments. Some studies have shown that exposure to low levels of certain toxicants may have a stimulating effect on growth. Indeed, it should be noted that an apparent impairment of one fitness component does not automatically imply a decrease in the overall fitness. It is therefore desirable to measure a number of parameters in addition to growth to provide a more accurate assessment of the effect of the toxicant involved.

Growth is usually estimated as an increase in body size or weight over time. Alternatively, other measures can also provide indirect information on the growth performance of individual organisms. For example, food conversion efficiency (FCE) of individual organisms in response to various levels of toxicant can be estimated. In addition, scope for growth (SfG), defined as the difference between energy intake and total metabolic losses, has also been used to measure the amount of energy available for production in an organism. SfG provides an indirect measure of the energy available for growth rather than a direct estimate of the growth rate.

FCE and SfG have been used in toxicological assessments of a wide range of sublethal stressors in both marine and freshwater environments.

Reproductive efforts and fecundity

Reproduction is one of the most sensitive endpoints for chronic or sublethal tests, and is biologically meaningful as it interferes directly with the production of future generations. Effects on reproduction may range from delayed sexual maturity, reduced fecundity, decreased percentage hatchability or hatching rate and lower offspring survival to complete breeding impairment. The effects may be temporary due to a short-term interruption of gametogenesis or more permanent due to cellular damage of reproductive tissues. Toxicological tests using reproduction as a criterion of effect, though costly, are useful and informative.

Development

An assessment of chronic toxicity should ideally involve an examination of the complete life cycle and/or multiple generations of test organisms. However, these elaborate tests are often too costly and time consuming and, thus, are often substituted by sublethal tests that focus on only one or two fitness-related response criteria. The assumption is that any adverse impact on the key fitness traits will have a subsequent deleterious effect on the state of health of the organism (see above). The effects of toxicants on the development of test organisms, particularly at the early life stages, have been widely studied. This may be a reflection of the general belief that early life stages may be more susceptible to toxic stress. Indeed, embryo–larval stages of marine organisms are often considered to be more sensitive to toxicants than adult stages and, consequently, criteria based on the developmental processes of early life stages are often employed in standard ecotoxicological assessments because of convenience and greater sensitivity.

Morphological change

Morphological deformities in organisms have been proposed as a biological screening tool to detect and assess the impact of environmental contaminants. Fish diseases such as fin erosion, skeletal deformities and tumours have been noted in contaminated waters. For example, fin rot, a condition involving progressive death of tissue and erosion of fins, has been observed in benthic fish exposed to contaminated sediments. Shell diseases, such as a brownish-black erosion of the exoskeleton and abnormal shell thickening, have also been reported in benthic crustaceans and molluscs. The development of male sex organs in female gastropods (a phenomenon referred to as imposex) is a well-documented example of morphological changes induced by the antifouling paint additive tributyltin (TBT). Indeed, indices based on morphological deformities in chironomid larvae have been developed and used to assess the long-term effects of chronic toxic stress in freshwater ecosystems. Obviously, different types and degrees of deformities need to be clearly and objectively defined and categorized so that they can be employed to gauge the severity of a specific response. It should be noted that morphological variations (deformity is a form of variation) may occur in natural healthy populations as a result of phenotypic or genetic differences, or both. The hypothesis here is that deformities in unstressed populations occur less frequently and are less severe than in those living under more stressed conditions, and that deformities are caused by the environmental stress.

Cytological or histopathological change

Cytological or histopathological changes are useful indicators of possible toxic effects that influence the normal structure and function of cells or organelles. Changes in the ultrastructure of cells or tissues, such as changes in the number, size and shape of specific organelles, particularly those related to protein synthesis and/or detoxification mechanisms, are useful diagnostic features of early cytological changes or damages resulting from toxicant exposure. Exposure to polycyclic aromatic hydrocarbons (PAHs) and algal toxins may contribute to tissue damage and tumour formation in fish liver. For example, exposure to organic pollutants such as polychlorinated biphenyls (PCBs) and PAHs may result in the proliferation of the endoplasmic reticulum systems and an increase in the abundance of lysosomes and autophagic vacuoles.

Biochemical change

Effects of toxicants on the biochemistry of organisms have traditionally received less attention than studies at higher levels of organization, such as individuals, populations, communities and ecosystems, mainly due to the fact that toxicologists are principally concerned with the well-being of individuals. Nevertheless, there is a recent move towards a greater use of biochemical markers in ecotoxicological studies. It is recognized that a better understanding of the biochemical changes following exposure to a toxicant can enhance our understanding of the potential toxic mode of action, and provide a solid theoretical foundation upon which predictions can be made regarding potential toxicities. This approach is particularly useful if the toxicant has a direct impact on biochemical pathways via its action on a specific enzyme or type of enzymes. One example is the commercially available Microtox® toxicity testing system which measures the light output of luminescent bacteria (*Photobacterium phosphoreum*) after they have been challenged by a sample of unknown toxicity. It compares the light output so produced with a control that contains no sample. The degree of light loss indicates a metabolic inhibition of the test organisms, and is used as an indication of the relative toxicity of the sample. With recent advancements in molecular biological and biochemical techniques, many sensitive biochemically based tools are now available which can indicate the exposure of particular biological systems to specific toxic stress.

Behavioural changes

The behaviour of an organism, integrating responses at nervous, muscular and energetic levels, has an important influence on its foraging efficiency, mating success, and ability to find shelter and avoid predators. These behaviours often have an important influence on the individual's fitness, and ultimately the survival of the species. Provided that organisms play a significant role in the functioning of their ecological system, the impairment of such behavioural responses will have an indirect effect on the ecosystem. For example, the phototactic response of planktonic organisms has been employed by some workers in developing sublethal bioassays for a number of xenobiotics. Changes in the light-induced orientative ability of marine larvae have also been used in rapid screening bioassays for ecotoxicological assessments. Avoidance behaviour has been widely studied, particularly in fish. It is useful to note that animals may avoid many kinds of toxicants or some generally unfavourable conditions and tests involving behavioural changes are often therefore nonspecific. Elaborate electronic devices have also been used to monitor the swimming behaviour (speed and pattern) of fish and the valve movement of clams exposed to xenobiotics. There is a need to standardize the testing procedure (e.g. test duration) to provide comparable data for use in a regulatory context.

Extrapolation of toxicity data on animals to other biota

Extrapolation from single species to multispecies

Tests involving single species are convenient and easy to perform. However, these tests deal with communities as if they were merely a collection of co-occurring species, and the behaviour of different species could be predicted by observing the response of a single indicator species. They cannot provide a general prediction of the impact of the toxicants on between-species inter-actions, which may be of importance under natural field conditions. Indeed, these testing designs fail to address the interactive processes that may modify the effects of toxicants. Important chemical factors may include biotransfor-mation, degradation and adsorption, while physical factors such as temper-ature fluctuations, dissolved oxygen content and water hardness and chemistry are also of relevance. The toxicity of certain chemicals may also be influenced by biological factors such as acclimatization, behaviour, resistance, predator–prey interactions, and intraspecific and interspecific competition.

Extrapolation from one species to other species

Test organisms (e.g. indicator organisms) are chosen on the assumption based on past experience, or biological intuition, that they are, to a certain extent, representative of the type of organisms that are likely to be exposed to the tox-icant. In other cases, organisms may be selected because they are particularly sensitive. Clearly, the type of information provided by the tests would vary depending on the selection criteria adopted. Even if carefully selected 'typical' and/or 'sensitive' organisms are used in the tests, chemicals can exhibit very different toxicities to different organisms, and an organism can be sensitive to one chemical but tolerant to another so extrapolation may not be justifiable.

Extrapolation from individuals to higher levels of biological organization

One of the major challenges in ecotoxicological studies concerns the difficul-ties involved in translating results and predictions from one level of biological organization to another, such as predicting population and/or community consequences from individual responses. One approach is to undertake toxic-ity tests with the most sensitive local species of economic, recreational or eco-logical importance. The assumption is that by protecting the most sensitive member of the community, the integrity of the ecological system will be pro-tected. Nevertheless, it is usually difficult to identify an organism which meets all the criteria. Thus, toxicity testing protocols have been expanded to include representatives of different trophic levels of a food chain. It is assumed that testing an array of sensitive organisms of different representative groups will improve the chance of covering the response range in the environment without assessing all possible species individually.

Extrapolation from laboratory to field conditions

Investigations on the ecological impact of pollutants are usually carried out in the field, while toxicological information is commonly obtained from laboratory studies based on a limited number of test organisms. Clearly, conditions in the field are different from conditions that can be created or simulated in a laboratory. As the toxicities of individual chemicals are likely to be modified by environmental factors (e.g. pH, water hardness, organic matter content), it is almost impossible to predict the precise effects of a chemical on an ecological system in a field situation on the basis of laboratory observations. It would be difficult to solve this problem completely but certain steps can be taken to alleviate it. One approach is to improve our understanding of the fate of chemicals and their mode of action on the relevant biochemical and physiological processes, particularly those directly related to the survival (such as growth, reproduction, energy utilization and osmoregulation) of the organisms or other biological systems in the anticipated receiving environment. A comparison of the toxicological effects of chemicals on the same selected species under both laboratory and field conditions will provide additional useful information on the appropriateness or validity of extrapolating toxicity data to effects in the field. Attempts have also been made to simulate environmental conditions inside the laboratory, i.e. construction of test microcosms and mesocosms. These model ecosystems tend to incorporate abiotic components with an array of organisms representing various functional groups or trophic levels of the biota or with samples of the community transplanted from field to laboratory. The relative merits of these approaches have been explored in Chapter 5.

Conclusions

Most toxicity tests currently used involve single-species laboratory tests. The effect criteria used include survival, growth, reproduction, physiology or biochemistry, behaviour or morphology, etc. but the majority of these tests are short-term and based on mortality. The test organisms are mainly invertebrates and fish. The dominance of single-species over multispecies, and acute over chronic tests, is attributable to the fact that the former procedures are relatively simple and easy to perform compared to the latter. Notwithstanding the usefulness of short-term, single-species tests, they are less effective in providing meaningful predictions as to the potential impact of toxicants on complex communities, and this may limit their use in environmental risk assessment. There is also a lack of consistency in comparing field- and laboratory-based test results for specific chemicals. It is clearly instructive to understand the toxic mechanisms involved in order to use available toxicity data for predictive studies. Overall, many of the approaches and procedures described in this chapter suffer from the problem of induction in that a specific effect caused by a specific chemical on a specific organism does not guarantee that other chemicals will have the same effect on the same system or that the same chemical will have the same effect on different organisms. Moreover, a demonstration of toxicity in a test situation does not automatically suggest any ecological significance.

The ultimate purpose of carrying out toxicological tests is to make use of particular observations (test results) to predict general environmental consequences. These predictions may include the potential impact on:

1 a particular species that needs to be protected for commercial, ecological or cultural reasons;

2 specific assemblages of organisms that need to be protected as they provide important ecological functions, such as energy transformation and nutrient recycling; or

3 ecological systems that need to be preserved for conservation purposes.

Traditionally, toxicological tests often rely on laboratory determinations of the mortality of test organisms exposed to varying concentrations of specific toxicants. These acute tests are usually fairly reproducible and relatively easy to perform, and have been incorporated into standard testing protocols in the US, Europe and elsewhere. These tests are important as a screening technique and a comparative tool. Notwithstanding their usefulness, acute tests have been criticized for their lack of ecological realism, and their inability to predict environmental effects of toxicants at sublethal concentrations. To counter this problem, complete life-cycle or multiple-generation chronic toxicity tests have been developed. However, the use of these tests has been somewhat limited due to the high demand for time and resources. Clearly, the challenge for ecotoxicologists is to develop chronic sublethal tests that can complement the existing acute tests and are, at the same time, easy to perform and of modest cost. There are response criteria that can provide an indirect measure of the status of health of test organisms. These criteria include key life-history parameters (e.g. growth rates and reproductive efforts) or patterns of energy allocation (e.g. scope for growth) of the organism.

Further reading

Duffus, J.H. (1980) *Environmental Toxicology*. Edward Arnold, London.

Rand, G.M. (1995) *Fundamentals of Aquatic Toxicology: Effects, Environmental Fate and Risk Assessment*, 2nd edn. Taylor & Francis, Washington, D.C.

Rombke, J. & Moltmann, J.F. (1996) *Applied Toxicology*. CRC/Lewis Publishers, Boca Raton.

Shaw, I.C. & Chadwick, J. (1998) *Principles of Environmental Toxicology*. Taylor & Francis, London.

7: Biomonitoring and Biomarkers of Hazards in the Environment

Introduction: principles of biomonitoring and biomarkers

Biomonitoring and the use of biomarkers, are two of the 'mainstays' of modern ecotoxicological assessment. Both refer to the use of living organisms for the assessment of environments, but the meanings of the terms are often confused. *Monitoring* refers to the assessment of environmental parameters on a regular basis in time. Thus, for example, the concentrations of various contaminants can be measured in waters, soils, sediments, air and biota frequently throughout a year—a practice commonly referred to as *environmental monitoring*. Monitoring can also refer to the assessment of various biological factors as well as purely chemical measurements. *Biomonitoring* is a sub-branch of monitoring, and specifically refers to the use of living organisms in monitoring procedures. Biomonitoring may take many forms—from the measurement of chemical residues in the tissues of living organisms through to the quantification of a number of biological endpoints including changes to various biochemical, physiological, morphological or behavioural factors, as well as traditional ecological measures such as the abundance and diversity of different species present in a community or an ecosystem.

Biomarkers, as we have seen in previous chapters, have become increasingly important in ecological assessments. The term *biomarkers* has been defined by the National Academy of Sciences in the US as follows: 'A biomarker is a xenobiotically induced variation in cellular or biochemical components or processes, structures, or functions that is measurable in a biological system or sample' (National Research Council 1987). Let us simplify the definition to its component terms. Biomarkers are *endpoints* of ecotoxicological tests which register a potential effect on a living organism. An endpoint is simply the response which is measured in a living organism during an ecotoxicological test. These tests are often conducted because of xenobiotic compounds in the environment but there can be other reasons. The term *xenobiotic compounds* refers to 'foreign compounds'—i.e. those substances which are not naturally present in, and have been formed by or introduced into the environment as a result of human activities. Thus, there are two major aspects related to biomonitoring and biomarkers—the presence of toxicants in the environment, and tests conducted on living organisms.

The various endpoints can be broadly divided into two categories, as follows.
1 *Physicochemical measurements*. These relate to the amount of a chemical present in a living organism. Such measurements can be related to concentrations

in abiotic sectors of the environment, e.g. in waters, soils, sediments and air, and to concentrations present in biota. The latter is especially important, as the uptake of contaminants by living organisms, i.e. the bioaccumulation potential, can be assessed. In addition, the amounts present in living organisms may be directly related to possible effects, if such a relationship has been previously established. Thus, the *degree* of any effect may not necessarily be realized through the use of biomarkers of exposure, but the *extent of exposure* can be assessed.

2 *Effects measurements.* These are biomarkers which actually demonstrate, through various endpoints, the effect of a particular chemical or a mixture, and its likely outcome on an individual organism, a population, a community or an ecosystem. By their very nature, these biomarkers can be used to predict what *may happen* if such chemicals are disposed of into a particular environment, or what will happen if exposure continues to occur.

Some biomarkers are specific. The results produced by such studies may be sufficient to replace chemical analysis of the surrounding environment, such is their sensitivity. Others—probably the majority—are much less sensitive, and only provide a generalized indication that a living organism may be suffering from stress induced by the presence of toxicants.

Many scientists view biomarkers merely as responses at the molecular, biochemical or physiological levels. Others take a wider perspective, and include the accumulation of chemicals in the tissues of living organisms, and responses which occur at the level of the individual, population, community or ecosystem.

Why do ecotoxicologists use biomonitoring and biomarkers? Simply because the responses measured are occurring in *living organisms,* and the information thus generated is particularly applicable to the management, protection and conservation of natural environments. The approach can be illustrated clearly by imagining a hypothetical relationship between the 'health status' of an organism and its response to increasing concentrations of contaminants in the environment (see Fig. 7.1). In a pristine environment, assuming all nutrients and food materials are available, an organism will be in a state of homeostasis, and can be classified as 'healthy'. As pollutant concentrations increase, the organism comes under greater stress until, at some point, it can be viewed as being 'diseased'. During the early phases of this process, the organism may be able to compensate in some fashion for the stress under which it is being placed. At some point, however, the organism can no longer compensate, although if the pollutant were removed from the environment, the organism could recover (i.e. the 'disease' is reversible). Eventually, however, when the pollutant insult is too severe, the organism reaches a point where the disease is no longer curable, and compensation is no longer possible (i.e. the situation becomes irreversible). Death may be the ultimate fate. Different biomarkers can be used throughout this continuum to assess the health status of the organism (see the lower graph of Fig. 7.1). It is desirable

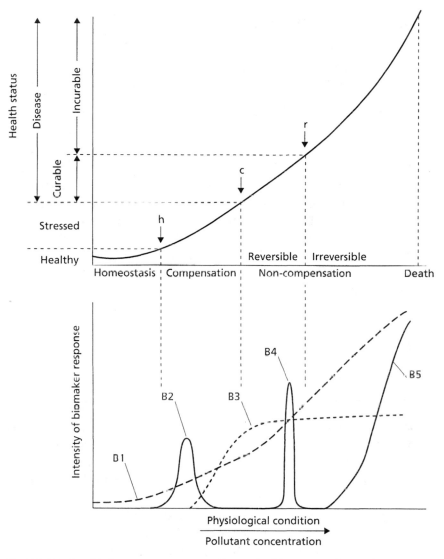

Fig. 7.1 Relationship between exposure to pollutant, health status and biomarker responses. The upper graph shows the progression of the health status of an individual as exposure to pollutant increases. h, The point at which departure from the normal homeostatic response range is initiated; c, the limit at which compensatory responses can prevent development of overt diseases; r, the limit beyond which the pathological damage is irreversible by repair mechanisms. The lower graph shows the response of five different hypothetical biomarkers used to assess the health of the individual. From Depledge *et. al.* (1993).

that these tests accurately reflect the differing status of the organism, thus providing a detailed picture of its health, and also reflecting the status of the environment in which the organism lives. It is important to note that more than one biomarker is likely to be needed to accurately monitor the situation.

Using living organisms as monitors

Early studies of chemicals in the environment revealed that many living organisms accumulated certain toxicants to levels much higher than those present in their ambient environments. The idea that such 'bioaccumulators' could be used in routine monitoring procedures was thus established. This was to lead to many monitoring programmes around the world utilizing so-called 'sentinel species' such as mussels and oysters as indicators of contaminants in marine and estuarine environments. But what was the relationship between the tissue concentrations of pollutants in living organisms and the toxic effects which they could cause? The widespread chemical monitoring of contaminants in living organisms soon became complemented by various toxicity tests which attempted to relate dose (i.e. ambient environmental concentrations or levels reached through bioaccumulation) to effect. The common effect measured in the early days of such testing was death. As time went on, however, more subtle endpoints, reflecting chronic exposure to contaminants, were employed—the beginnings of what we would now recognize as *biomarkers*.

Today, we also recognize the complexities in relating biomarker evaluations at the subcellular, cellular and individual organism level to effects on populations and ecosystems. At times, a 'quantum leap' is involved in making predictions for higher levels of environmental organization based solely on chemical or biomarker data, which have been obtained using lower levels of organization.

A classification of biomonitoring types is based on the way in which contaminants enter the environment and how they subsequently enter and affect living organisms. This involves: the *initial exposure* of living organisms to contaminants and the consequent *uptake* of various compounds in their tissues; the *effects* which may be caused within cells, tissues and individual organisms (including genetic effects); and the effects at higher levels of organization, such as populations and communities. Thus a classification is as follows:

1 biomonitoring of bioaccumulation of contaminants;

2 sub-cellular, cellular and individual organism biomarkers (including those at the genetic level); and

3 biomonitoring of ecological effects of contaminants.

Throughout this chapter, we shall also bear in mind some other important facts which Hopkin (1993) has called the 'five Rs' of biomonitoring. These are rules based upon the use of biomonitoring techniques in assessing contaminants in field situations, and can be considered to be guidelines for the successful assessment of the effects of pollution on living organisms. The 'five Rs', Hopkin believes, should be satisfied in any *in situ* usage of biomonitors for the practical assessment of the effects of pollution. Specifically, they include:

1 *Relevance*. Any biomonitoring programme should be relevant to the environment under consideration. Thus, for example, the species involved should be those present (and important) within the environment.

2 *Reliability*. If tests are to be reliable, then the species involved should be widely distributed in an environment, and they should be commonly found in order to allow accurate comparisons to be made between one locality and another, especially when these localities are separated by a great distance.

3 *Robustness*. For many types of biomonitoring, it is desirable that the organisms involved should not be killed by very low levels of contaminants. This is especially true of bioaccumulators (i.e. sentinel organisms such as mussels), where the level of contaminants in their tissues is the endpoint being measured. If such organisms were to die on exposure to low levels of contaminants, then the tests would not be practical. The organisms used in these sorts of experiments should ideally be capable of being transplanted into a new test locality and of surviving throughout the period of experimentation.

There are, however, exceptions to the rule of 'robustness'. For example, particularly sensitive species within communities are the first to be affected by pollutants, and their loss (through death) may be a sensitive measure of the effects of contaminants on this level of organization.

4 *Responsiveness*. Responses that biomonitoring organisms exhibit should, by definition, be measurable, and the bigger the toxicant insult, the greater should be the response in the measurable endpoint. Thus, ideally, there should be a clear dose–response relationship involved.

5 *Reproducibility*. The same response should be elicited by a test organism wherever it is appropriately used. The response should also be consistent with the amount of the environmental exposure (i.e. the dose) which has been applied.

The occurrence of chemical residues in organisms: some basic principles

Bioaccumulation of contaminants by living organisms can occur through two routes (see Phillips 1993 for a comprehensive review):

1 *bioconcentration*, which may be defined as the accumulation and sequestration of contaminant materials by living organisms directly from the ambient environment; and

2 *biomagnification*, which refers to the accumulation of contaminant materials via the dual processes of bioconcentration and trophic transport. In this case, contaminants are accumulated primarily through food chains or food webs, and concentrations increase with trophic level (i.e. towards higher consumers).

Bioaccumulation of contaminants depends upon a number of physicochemical factors (such as chemical speciation, partitioning and degradability); biological variables (including the species involved, its habitat, physiology, feeding habits, etc.); and environmental factors (including season and local hydrodynamics) that may alter the distribution and bioavailability of individual contaminants.

Table 7.1 The periodic table of elements, showing those metals that have physiological functions.

Na[a]	Mg[c]										
K[b]	Ca[c]			V[f]	Cr[f]	Mn[d,e]	Fe[d]	Co[f]	Ni[f]	Cu[d]	Zn[e]
					Mo[d]						

a, Components of chemical entities; b, osmoregulators, charge balance, action potentials; c, control functions, structure factors, enzyme cofactors; d, electron transfer reactions, oxygen carriers; e, acid–base catalysts, structure factors; f, beneficial to some organisms at low levels.

There are three main implications of bioaccumulation, which make it important in biomonitoring activities.

1 Even though environmental concentrations of contaminants may be low, concentrations in an organism's tissues may reach the point where they are toxic.

2 Even if such concentrations do not affect the organism itself, they may adversely affect the health of predators, even including man.

3 The concentration in the tissues of some organisms may be used as an indicator of concentrations present in the surrounding environment. This concept has been adopted worldwide, and is the driving force of programmemes such as the 'Musselwatch', which uses bivalve shellfish as 'indicator' organisms to assess levels of organic and trace metal pollution.

Living organisms use differing strategies to accumulate contaminants from the environment. In the case of trace metals, for instance, some living organisms use what can be called 'evolutionary strategies', thus satisfying their physiological requirement for certain 'essential' metals (see Table 7.1) which are often employed in enzymes or cofactors. In addition, these organisms may detoxify 'nonessential' metals which have been accumulated from the environment or from food.

Most metals are accumulated from the environment via passive processes, along concentration gradients. However, in the case of certain metals such as cadmium, uptake may also be linked to active uptake mechanisms, which transport other essential metals across biological membranes. Terrestrial organisms tend to be more efficient in their selective uptake and exclusion of trace metals than do aquatic species, perhaps as a result of their less frequent natural exposure to high metal concentrations.

As trace metals can be toxic, living organisms have also evolved strategies to counter their effects. The main tactics employed here are: regulatory processes, which result in a reduction in metal uptake or an increase in metal excretion; and sequestration ('hoarding') strategies, which result in the

Non-regulators   Extreme uptake

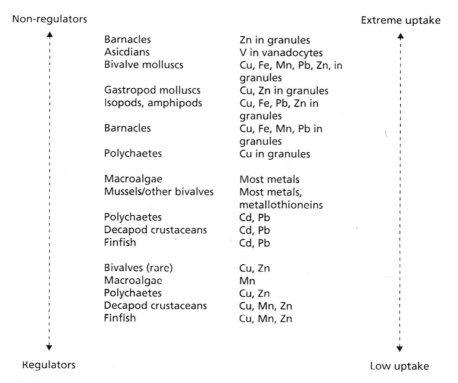

Barnacles	Zn in granules
Asicdians	V in vanadocytes
Bivalve molluscs	Cu, Fe, Mn, Pb, Zn, in granules
Gastropod molluscs	Cu, Zn in granules
Isopods, amphipods	Cu, Fe, Pb, Zn in granules
Barnacles	Cu, Fe, Mn, Pb in granules
Polychaetes	Cu in granules
Macroalgae	Most metals
Mussels/other bivalves	Most metals, metallothioneins
Polychaetes	Cd, Pb
Decapod crustaceans	Cd, Pb
Finfish	Cd, Pb
Bivalves (rare)	Cu, Zn
Macroalgae	Mn
Polychaetes	Cu, Zn
Decapod crustaceans	Cu, Mn, Zn
Finfish	Cu, Mn, Zn

Regulators   Low uptake

Fig. 7.2 Examples of organisms that employ different strategies for the uptake, sequestration and excretion of trace metals. After Phillips & Rainbow (1989).

accumulation of large amounts of some elements within the tissues of the organism. Aquatic species can be considered to lie on a continuum between 'regulators' (showing low net uptake of metals due to efficient exclusion or excretion processes) and 'nonregulators' (which tend to accumulate large concentrations of metals) as shown in Fig. 7.2.

In the case of organic contaminants, quite different mechanisms of uptake are seen. Many organic contaminants are lipophilic (i.e. poorly soluble in water, but readily soluble in fats). Lipophilic contaminants include the chloro-hydrocarbons such as DDT, the polychlorinated biphenyls (PCBs), petroleum hydrocarbons and polycyclic aromatic hydrocarbons (PAHs). Bioaccumulation of these contaminants depends upon physicochemical processes based on the partitioning of the contaminants between an aqueous phase (in the environment) and a nonaqueous phase (in fats in the tissues of living organisms). The K_{OW} values (see Chapter 3) are very useful indicators of bioaccumulation for these organic compounds. Food may also make an important contribution to the accumulation of lipophilic compounds, especially in the case of larger aquatic vertebrates, marine mammals and most terrestrial biota (see Table 7.2).

Table 7.2 Model system for the bioaccumulation of lipophilic contaminants. The importance of routes of uptake and loss are indicated on an increasing scale from (+) to (+++++). After Walker *et al.* (1996) and Phillips (1993).

Type of organism	Routes of uptake			Routes of loss	
	Diffusion	Food	Water	Diffusion	Metabolism
Aquatic					
Molluses	+++++	+		+++++	
fish	++++	+ to +++		++++	+ to +++
Terrestrial					
Most animals		++++	+		+++++

The lipid content of an organism is the most important factor affecting organic contaminant uptake, and determines differences in bioaccumulation. For example, differences in lipid content may be observed between individuals of a species, depending upon age and sexual condition. This has been particularly well demonstrated in marine mammals, where variations in chlorohydrocarbon concentrations are correlated with age, sex and lactation. In addition, within an individual organism, differences in tissue lipid content or differences in the type of lipids in various tissues may also influence uptake. For example phospholipids, which are commonly found in brain tissue, are polar in nature and as such have a lower capacity for the accumulation of lipophilic compounds.

Biotransformations of contaminants are also important in determining the residence time of contaminants—both metals and organics—within organisms. Trace metals can be transformed biologically in various ways, and various metal species may be accumulated to differing degrees by living organisms. The best-recognized transformations of metals are those involving formation of organometallic compounds, such methyl mercury. The methylated forms of mercury (Hg) produced by bacterial activity are highly toxic, and can accumulate to high concentrations in living organisms.

Most of the higher organisms are capable of transforming persistent organic compounds through the use of hepatic microsomal monooxygenase systems which are inducible in nature and utilize cytochrome P450. Biotransformations involving chlorohydrocarbons and petroleum hydrocarbons involve two stages: oxidation, hydrolysis or reduction of the parent compound; and the production of excretable conjugated metabolites. However, biotransformation ability varies considerably between species, and as a general rule is more advanced in terrestrial biota and aquatic vertebrates than in aquatic invertebrates. Compounds tend to be more resistant to transformations when they do not have functional groups (such as ester, hydroxy, nitro or amide groups), and are especially resistant in cases where carbon–hydrogen (C–H) bonds are protected from attack by high levels of halogenation. Thus, certain PCB congeners (the non*ortho* coplanar PCBs, which are essentially analogues of 2,3,7,8-tetrachlorodibenzo(*p*)dioxin) are very resistant to biotransformation, and are relatively toxic.

There are several other important factors that influence bioaccumulation of the trace metals. These factors include salinity, temperature, pH, oxygenation and contaminant interactions with other contaminants. These factors principally operate by causing changes in chemical speciation of trace metals or by altering partitioning between dissolved and particulate phases.

Using bioaccumulation in monitoring

Many organisms have been used as indicators of bioaccumulation, from algae to bivalves, and from birds to mammals, including man. Historically, the results have been interpreted not only as an indication of exposure, but also as an indication of potential adverse effects. The usual goals of such programmes were the provision of long-term monitoring data on the concentrations of contaminants, and the identification of localities or species where accumulation occurred in greater than anticipated concentrations. This was an attempt to provide an 'early warning' system that a particular region or organism might be subject to pollution, and that remedial action might be required.

The bivalve shellfish, such as oysters and mussels, have perhaps attracted the most attention as monitors used in this fashion, and as a result epitomize the basic characteristics of bioaccumulator organisms used in routine monitoring activities. Many bivalves are commercially important species, and the contaminant concentrations in their tissues are 'integrated' over both time and space. As bivalve shellfish are filter feeders, they readily accumulate contaminants; as they are sedentary, they represent contaminant availability within the area they live. As they can be harvested regularly, they also provide information on changes in contaminant concentrations over time.

Of course the contaminants are much more easily measured in bivalves than in water, where they exist in much lower concentrations. They also represent that fraction of contamination in the ambient environment which is 'biologically available'. Bivalve shellfish are also widespread throughout the globe and, in the case of the organic contaminants, they have a low ability to metabolize xenobiotic compounds, thus more accurately predicting the presence of these substances.

However, organisms such as the bivalve shellfish also suffer from disadvantages as monitors. For example, in the case of trace metals and some other substances, there are long equilibration times (i.e. the time taken to reach a

'steady state' in tissue concentration on exposure to environmental concentrations). Site-to-site variations in the species can also prejudice results
(including factors such as sex, size, life-cycle status, etc.). Physical differences
in the sites where the organisms are collected can also cause difficulties in
interpretation of results—for instance, differences in temperature, salinity, etc.
Despite these difficulties, many environmental protection organizations rely
on data collected from bioaccumulator species as an integral part of management practices. The range of chemicals monitored is usually limited however;
the major contaminants—trace metals (in particular Zn, Cu, Cd, Pb, Hg), the
pesticides (especially the chlorohydrocarbons such as DDT) and PCBs—
are amongst the most common pollutants on any monitoring 'short list'.
Somewhat less frequently, petroleum hydrocarbons and PAHs are monitored.
Even less frequently (and usually subject to the peculiar circumstances that
may be involved) radionuclides may also be included in some monitoring
programmes.

The choice of organisms used in bioaccumulator monitoring programmes
has been the subject of considerable research. The attributes desirable in these
organisms are as follows.

1 The organisms should accumulate the pollutant without being killed by the
environmental levels present.

2 The organisms should be sedentary in order to be representative of the area
in which they are growing.

3 The organisms should be readily found throughout the area being monitored.

4 The organisms should be relatively long lived.

5 The organisms should be of a reasonable size to ensure that enough material
is available for analysis.

6 The organisms should be easy to handle, both in the field and in laboratory
situations.

7 The organisms should be able to tolerate brackish water, where pollutants
are frequently found.

8 There should be a simple correlation between the concentration of pollutants in the environment, and the levels found in the tissues of the living
organisms.

The bivalve shellfish, such as mussels and oysters, fulfil most of these characteristics, which is the main reason why they have become widely useful
sentinel indicator organisms. In some countries, they have been widely
used in monitoring programmes, which have become popularly known as
'Musselwatch'. For instance, the California State Musselwatch has undertaken studies of trace metal and pesticide distribution in Californian waters
annually since 1977 (see Box 7.1).

Despite the popularity of measuring contaminants in molluscs such as the
bivalve shellfish, a huge variety of other organisms have also been utilized.
These include phytoplankton, macroalgae, ascidians, annelids, crustaceans,
finfish, birds and mammals. Monitoring has not been restricted to marine

Box 7.1: The Musselwatch

Programmes employing bivalve shellfish as sentinel monitors of conservative contaminants (including trace metals, pesticides and PCBs) have been undertaken in many parts of the world. Amongst the earliest efforts were those undertaken in the US during the mid-1960s, which eventually resulted in the establishment of the National Monitoring Programme. Many other countries have also routinely used bivalve shellfish as monitoring organisms, including the UK, several Mediterranean countries, Hong Kong and Australia. By far the most popular organism used has been the humble mussel—and hence, the term 'Musselwatch' was coined.

The Musselwatch technique relies upon the ability of shellfish (e.g. *Mytilus edulis*) to accumulate contaminants in their tissues far above the concentrations found in the surrounding environment. They also have the ability to 'integrate' the concentrations to which they have been exposed—in other words, they reflect the environmental concentrations present in the ambient environment over a period of time.

One of the longest running Musselwatch programmes in the world has been conducted in California (the California State Musselwatch), where it is sponsored by the California State Water Resources Control Board, and operated by the California Department of Fish and Game. The goal of this programme is to provide an indication of spatial and temporal trends of contaminants in nearshore waters along the state's coastline. Thus, the data can be used to show regions where problems exist (i.e. high concentrations of various contaminants are found in the mussels), and how these concentrations vary over time. The mussels which are used are all taken from a clean site, and transplanted to the Musselwatch selected sites once per year.

In the early days of the programme, research was undertaken to ensure that the study design considerations had been met, and that the data could be interpreted successfully. Thus, suitable analytical techniques were developed, appropriate species were selected, field sampling techniques were tested, and the kinetics of uptake and depuration within the mussels determined. Other factors were also taken into consideration in this study, including individual variability due to sex, age, tissue type, etc., and temporal variability due to seasonal or environmental fluctuations.

The California Musselwatch has successfully provided detailed information on the temporal and spatial trends of a host of contaminants in the 20 years since it was established. For instance, PCB and p,p'DDE distribution patterns along the California coast were determined, and temporal trends evaluated. In addition, certain 'new' compounds were identified as accumulating in mussels (the polychlorinated terphenyls).

habitats, with much valuable information being obtained about freshwaters and terrestrial environments (e.g. the role of monitoring birdlife in elucidating the role of DDT in eggshell thinning). Humans, too have been monitored— both in industrial settings (to obtain information on exposure to contaminants in the workplace) and following more general exposure (e.g. the analysis of lead in children's teeth, and of PCBs and chlorohydrocarbon pesticides in the breast milk of feeding mothers).

Bioaccumulator monitoring programmes are most often undertaken for two main purposes:

1 to establish spatial and temporal differences in contaminant distribution; and

2 to measure contaminants in organisms consumed as food to establish if acceptable health limits have been exceeded.

However, these two objectives are frequently confused in monitoring pro-grammes, and interpreting data accumulated for the latter objective in terms of spatial and temporal distribution is often fraught with difficulties. Many of the food species tested are mobile, and may not necessarily indicate that high concentrations are the result of contamination in the capture locality. In addi-tion, meaningful comparisons between different species of organisms and between the different analytical techniques utilized by various workers are difficult to draw. These difficulties highlight the need for monitoring pro-grammes with firmly established aims and methodologies if data are to be obtained which accurately indicate the extent of contamination. Laboratory-based protocols for the assessment of bioaccumulation in aquatic environ-ments have been developed by the Organization for Economic Cooperation and Development (OECD), the American Society for Testing and Materials (ASTM), and the US Environment Protection Agency (USEPA). Transplanted organisms, rather than indigenous stock, have often been used, thus cutting down on the inherent biological variability involved.

Use of biota to monitor hazards

The aim of using biomarkers is to relate toxic chemical presences in the envir-onment with effects on living organisms. The hazard which chemicals pose to organisms is related to the toxicity of the chemical (or combinations of chem-icals) involved, and the degree to which the organism has been exposed (i.e. the dose to which it has been subjected over a period of time). The net result of exposure and toxicity is an effect (i.e. an endpoint) which is measurable in some way, often by use of a biomarker.

Biomarkers range from highly specific responses to particular chemicals, to responses which are nonspecific and can be caused by a range of different substances in the environment. One of the best-known specific responses involves the enzyme aminolevulinic acid dehydratase (ALAD), which is an important enzyme in the biochemical pathway for the synthesis of haem. It is specifically inhibited by lead. So specific is this inhibition that information derived from ALAD assays can be used to predict exposure to, and the hazard represented by, bioavailable lead in the environment.

Although such a specific response provides valuable information, other relatively nonspecific biomarkers are also worthwhile contributors to the assessment of environmental exposure and its possible effects. For instance, the mixed function oxygenases (MFOs), otherwise known as monooxygenases, are enzymes which are induced by the presence of various chlorohydrocarbons, organophosphorous compounds and the PAHs. These enzymes—collectively known as the cytochromes P450—are important as detoxifiers of chemicals (both natural and foreign) to which organisms are exposed, and their induction is widely used as a biomarker. Although induction of these enzymes does not provide information on the individual chemicals which may be involved, it does provide a useful index of exposure, and indicates that the organisms concerned may be potentially affected by the presence of these chemicals.

Highly specific biomarkers are fewer in number than those which are relatively nonspecific. The corollary to this, of course, is that no one biomarker can provide all the necessary information on exposure to, or the effects of, a range of different chemicals present in an environment. Thus, in most environmental assessments, a suite of different biomarkers is used for monitoring purposes. It is important that, in choosing such a suite, no one organism or trophic level is targeted, and that the information provided allows an assessment of hazard to be made for the total area, community or ecosystem involved. There are biomarkers—both specific and nonspecific—at various levels of organization, commencing with the subcellular, cellular and whole-organism tests, and progressing through to measurements performed at the levels of population, community and ecosystem.

Biomonitoring at the level of the cell and the individual

Effects may be relatively easily measured at the level of the cell or the individual organism. Thus, for example, scientists have for many years used morphological or histopathological changes as indicators of exposure. These are usually gross changes in the structure of cells, tissues, organs or individual organisms which are relatively easily seen and measured. One such well-established change is the development of 'imposex', which has been observed in many areas of the world. Imposex is a response to the presence of tributyltin (TBT) compounds, which have been widely used as antifouling substances which prevent the growth of organisms such as barnacles on the hulls of ships. TBT causes changes to many organisms, for example alterations to shell structure and growth in oysters, but the development of imposex in molluscs such as the English dog whelk *Nucella lapillus* has been widely used as a biomarker of effect. Imposex results in female organisms growing a penis and this

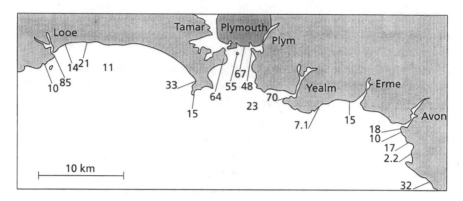

Fig. 7.3 Levels of 'imposex' (percentage size of female penis relative to males) in dog whelks (*Nucella lapillus*) collected from SW England in 1984–1985. Imposex develops in females in response to tributyltin (TBT) leached from antifouling paints and is most prevalent in areas of high boating activity (e.g. the Looe and Yealm estuaries). Reproduced from Bryan *et. al.* (1986) by permission of the Marine Biological Association of the United Kingdom and Cambridge University Press.

phenomenon can ultimately cause population failure as effective reproduction ceases. The biomarker measured in this case (the 'imposex index') is essentially a ratio of the mean female and the mean male penis size, and it provides an indication of exposure to TBT (see Fig. 7.3).

In recent years, there has been a rapid development of biomarkers at the subcellular level of organization, due not only to advances in biochemistry but also to modern methods of measurement which are rapid and cheap and can be performed in large numbers at one time. Such biomarkers include ALAD and the monooxygenase systems reported above, and also many others, so many that it is impossible to cover them all in this chapter (but see also Chapter 4 for a review). We shall concentrate on only a few examples which have proved popular in environmental assessment of the hazards of toxic chemicals to living organisms.

We have seen above that ALAD inhibition is an example of a specific biomarker for environmental exposure to lead. It has been widely used for the assessment of environmental exposure, not only in wildlife, but also in humans where it has become a standard method. In terms of monitoring wildlife exposure to lead, ALAD inhibition has been used in fish, rodents and birds (especially waterfowl which may be accidentally exposed to lead shot during hunting seasons).

Another specific response is used to assess exposure to organophosphorus and carbamate pesticides in the environment. This test involves the enzyme anticholinesterase (AChE), which is inhibited by these pesticides such that it cannot hydrolyse its substrate, acetylcholine. This is an important biomarker, as the organophosphorus and carbamate pesticides are notoriously difficult and expensive to analyse chemically, combined with the fact that they disappear rapidly from the body. These pesticides cause a variety of symptoms in living organisms which result from disturbances to nerve function, and the bioassay involving AChE has been shown to be closely related to the development of these symptoms.

MFO induction has been widely used in many parts of the world, but its success has been particularly shown in waters exposed to the waste products of paper manufacture. The complex halogenated mixtures produced by pulp mills in this process induce the monooxygenase enzyme system, and this induction can be measured via several tests, including the induction of ethoxyresorufin O-deethylase (EROD). Excellent results have been obtained using this method in such countries as Sweden, the US and Canada. However, it is important to realize that the MFO system is relatively nonspecific. Many organisms possess this enzyme complex, whose role is detoxification. As such, its induction is a response to a wide variety of compounds in the environment —including both natural compounds and xenobiotics. Thus, MFO results obtained in the field are often difficult to interpret, especially in localities which are not receiving point source inputs of chemicals that induce MFO activity.

A rapidly growing area of subcellular biomarker tests involves the interaction of the cell's genetic material (DNA) with toxic chemicals. These tests rely on the premise that any changes to DNA may have long-lasting and profound consequences. However, it is also important to realize that DNA does have a self-repairing capability, and thus results of many assays of genetic material may overstate the seriousness of the situation. Nonetheless, biomarkers involving genetic material are important indicators of possible problems in reproduction or possible carcinogenic effects. Measurements of DNA adducts, for instance, have become widely used as indicators of exposure to known carcinogenic PAHs, such as benzo[a]pyrene.

Aside from the subcellular biomarkers mentioned above, there are countless others. These include changes in the levels of key molecular components of the cell that can be used as an indication that an organism is under environmental stress. Such measurements include assessments of cholesterol and pyruvate concentrations and various changes to proteins. Of the latter, metallothioneins have been used as an indication of trace metal exposure. These proteins have a high affinity for certain metals (e.g. Cd, Cu, Zn, Hg), and their induction has been demonstrated in laboratory trials. Field assessments of metallothionein production have also been undertaken, but their usefulness as a biomarker is limited by a lack of understanding of a dose–response relationship, combined with the fact that there is considerable variation in production between different organisms and species. Certain proteins which are inducible—such as stress proteins (see Chapter 4)—also suffer from the fact that they respond not only to anthropogenic pollution, but also to natural phenomena. This makes them extremely difficult to use as 'stand-alone' biomarkers.

Relative concentrations of amino acids have also been used as a biomarker —for instance, the relative concentrations of taurine to glycine, and the combined concentrations of threonine and serine. The latter has been shown to be a sensitive index of stress in bivalve shellfish such as mussels.

Adenylate energy charge (AEC) provides an estimate of the energy potential which is available to an organism. This is calculated from the formula:

$$AEC = \frac{ATP + \frac{1}{2}ADP}{ATP + ADP + AMP}$$

and its value lies between 0 and 1. At a value of 0, AMP is the only component present; at a value of 1, all energy-related compounds are in the form of ATP. This index has been found to be particularly useful for determining stress in polluted aquatic environments.

Various physiological responses at the level of the whole organism have also been measured and utilized as biomarkers. These include studies of such basic physiological functions as respiration, changes in growth rate, feeding, excretion, etc. Many provide a good assessment of stress in organisms exposed to pollution, although it needs to be remembered that changes of this sort can also be caused by natural processes in the environment which cause stress (e.g. changes in temperature, oxygenation and salinity). Thus, relating the biomarkers to man-made pollution can be fraught with difficulty.

As physiological responses are subject to considerable natural variations, attempts have been made to make 'integrated' measures of an organism's well-being, based on a range of different functional parameters. This is the essence of 'scope for growth', which has been successfully used in freshwater, estuarine and marine situations.

Simpler measures of physiological stress caused by pollution can of course be performed. In many studies involving bivalve shellfish, a 'condition index' has been calculated to provide a simple means of telling whether the organisms involved are in a healthy or stressed state. In bivalve shellfish, a simple condition index involves calculating the proportion of the shell volume which is occupied by the flesh of the organism (i.e. shell volume divided by meat weight). Although crude, such a ratio can allow comparisons to be made between populations living under different conditions. However, as with many other physiological biomarkers, the natural effects of temperature, availability of food, life-cycle changes, etc. will also alter the condition index and make interpretation difficult. A more refined condition index involves the calculation of the O:N molar ratio—the ratio between the oxygen consumed by an organism and the nitrogen excreted. In essence, this is an estimate of the use by the organism of proteins in energy metabolism. A high usage of proteins (relative to carbohydrates) is often a sign of stress, and results in a low O:N ratio.

Physiological biomarkers of pollution stress are often subject to variability and changes induced by natural environmental effects. Thus, no one biomarker should be relied upon in any assessment of a polluted area. Combinations of various biomarkers, along with chemical data, are therefore frequently used in environmental monitoring of chemical hazards. A good example of this is shown in Fig. 7.4, which compares various data (e.g. chemical measurements, scope for growth, O:N ratio) in Narragansett Bay, USA, allowing an overall assessment of the area to be made.

Apart from physiological biomarkers used with individual organisms, behavioural responses can also be measured. Again, a myriad of different measurements can be undertaken, including sensory responses (such as phototaxis, chemotaxis, temperature preferences, tactile inhibition), rhythmic activities, motor activities, motivation and learning phenomena and interindividual responses such as migration, aggression and predation vulnerability. Behavioural responses can be viewed as representing a higher level of organization, as the endpoints measured are those which are 'integrated' at the whole organism level. However, these biomarkers are, by definition, difficult, tedious and often expensive to measure (especially in the field situation).

Among the most advanced of behavioural responses used as biomarkers are those observed in fish, involving avoidance behaviour. In this instance, well-established procedures are available for laboratory tests, and results are capable of accurately predicting effects under regimented protocols. Nonetheless, prior conditioning of the test organisms to contaminated waters can substantially reduce the effectiveness of the tests and result in 'desensitization' of the organisms. This highlights the difficulties involved in behavioural biomarkers. Many tests are simply not sensitive enough to provide meaningful conclusions, and are often much more costly and difficult to perform than tests at the biochemical or physiological level. This is why subcellular-based tests (in particular) have become so popular in ecotoxicology. They are (more often than not) rapid to perform; they are reproducible in the laboratory context; the accuracy and precision of the data generated is relatively high; they are sensitive to changes induced by pollutants; and often they provide results specific for particular contaminants (or groups of contaminants). However, do they represent pollution-induced changes which occur at higher levels of organization? A partial answer can be obtained from examining biomarkers at the levels of population and community (see later in this chapter).

Monitoring effects of hazards at the population and community levels

The major question in ecotoxicology is: can the more frequently measured parameters of disturbance at the lower levels of organization (e.g. chemical analyses involving bioaccumulator organisms and biomarker measurements) be related to changes at the population and community levels? Effects observed at the level of populations are usually related to changes in the number of individuals present, the reproductive output of a population, or the

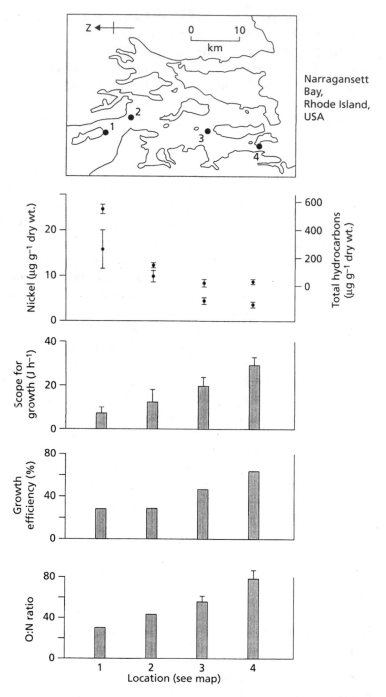

Fig. 7.4 Concentrations of nickel (solid circles) and total hydrocarbons (open circles; µg g⁻¹ dry weight) in the whole soft parts of mussels (*Mytilus edulis*) transplanted along a gradient of pollution in Narragansett Bay, USA (see map), and changes in scope for growth of the mussels, growth efficiency and oxygen : nitrogen ratios. After Widdows (1985).

recruitment of the organisms involved. Of course, it is difficult to separate changes at the population level from those at the level of communities, as changes in populations will by definition lead to changes in communities.

Some species are sensitive to pollutants in the environment, and their numbers decrease rapidly as a result of exposure to toxicants. Others can be classified as 'opportunists'—i.e. species which exploit a polluted situation due to their tolerance, or due to a lack of competition which has been caused by decreased numbers of other species. Thus, both these effects can be used as indices of pollution effects.

These types of changes are frequently seen where organic pollutants (e.g. sewage discharges) affect localized areas. For instance, in the marine situation, the polychaete *Capitella capitata* is a frequent colonizer of areas affected by sewage, as it is a tolerant species. It is able to reproduce rapidly and it has a high level of tolerance to organic pollutants, and as a result, its numbers increase vastly in the absence of any competition.

The dominance of populations such as *Capitella capitata*, which reproduce quickly and mature rapidly (the so-called 'r' strategists), is common in disturbed environments. This phenomenon has been observed around North Sea oil platforms, near sewage outfalls in southern Australia, in polluted rivers in England and in many other parts of the world. However, for those species with a longer life span and slower rates of reproduction (the 'k' strategists), a lowering of numbers is the most common response to contaminated environments.

Shifts in the age structure of populations have often been used as an indication of stress, based upon the premise that younger organisms (particularly at larval stages) are more sensitive to contaminants than adults. However, to this needs to be added the fact that, as populations age naturally, older organisms dominate, and sifting the effects of pollutants from natural effects becomes a difficult problem. Another strategy adopted to overcome these difficulties has been the examination of genetic effects at the population level. Such changes would be expected by differential lethality to organisms in a population, and the frequency of various genetic 'marker alleles' has been used as evidence for changes.

Bacteria can also be used as evidence of these effects—most beneficially because of their rapid growth rate under laboratory test conditions. For example, where various organic contaminants such as PAHs are present, the population of bacteria with an ability to degrade these compounds increases. Measurement of population numbers with such an ability in test site samples, in comparison with uncontaminated control samples, provides evidence of pollution. However, it is important to realize that the observed changes measured at the phenotypic level may not provide evidence for profound and longlasting genotypic changes, especially in organisms such as the prokaryotes which exhibit a degree of 'genetic plasticity', i.e. genetic adaptability, most likely via plasmid involvement, which may occur rapidly upon exposure to different compounds.

Changes occurring in populations have inevitable repercussions upon communities. A fall in population numbers—or the loss of a population—opens new niches for other species to fill. A rise in population numbers signals greater competition, which may have profound effects on other species in a community. Monitoring communities is therefore one of the most important aspects of assessing the impact of contamination on environments, and is a task which has been widely undertaken in terrestrial, freshwater, estuarine and marine environments.

The response most frequently measured at the community level is change in species abundance. These changes may occur over long or short periods, and the patterns involved may be marked or subtle. Some of the earliest convincing evidence that human-induced emissions to the environment could change community structure was recorded in lichen populations in England. Lichens are very sensitive to many air pollutants—in particular, sulphur dioxide (SO_2). In the UK, surveys of lichen populations have shown marked changes in response to air contamination (for example, see Fig. 7.5), which may be indicated by changes in the total numbers of species of lichens present, or alterations in species abundance. Monitoring of this nature has been continuing for many years, producing excellent baseline and reference data, and marked improvements in lichen numbers and species abundance have been seen in association with new laws that restrict the level of atmospheric contaminants from industrial and municipal inputs.

Many ecological studies have been undertaken in aquatic environments to assess the impacts of toxic chemicals; many different types of communities—from planktonic to benthic—have been investigated and various fisheries have

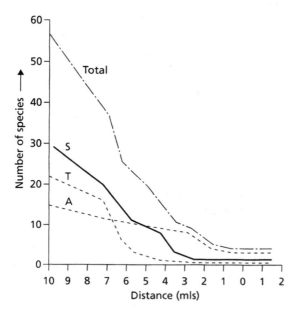

Fig. 7.5 Numbers of species occurring on asbestos–cement (A), sandstone (S) and trees (T) along a transect through the centre of Newcastle upon Tyne. From Gilbert, O. (1965) *Symposia of the British Ecological Society* **5**, 35–47.

been assessed. Often, however, the establishment of cause and effect has proven to be difficult. In some instances, defined measures have been adopted for assessment which have provided excellent data. Such is the case in the UK and in South Africa, where investigations of freshwater communities have been undertaken. In the UK, three major approaches have been adopted.

1 A biotic approach (which relies upon differences in the sensitivity of pollutants to certain species). Thus, the disappearance of sensitive species or an increase in tolerant species is used as an index of contamination.

2 A diversity-based approach, relying on changes in the diversity of a wide variety of organisms present in a community.

3 The *River Invertebrate Predation and Classification* system (RIVPACS), combining investigations into the types of species present and their abundance.

In estuarine and marine situations, the presence of fouling organisms has been utilized, as data are relatively easily obtained in this area. For instance, fouling panels may be placed in various localities (at both contaminated and reference sites) and the range of organisms developing on these panels can be assessed. In estuarine and marine situations, the presence, diversity and abundance of benthic organisms has also been very widely used as a successful measure of the effects of contaminants. For example, oil platforms in the North Sea have been extensively investigated using various indices of benthic faunal presence (see Fig. 7.6). These have shown an increase in the diversity indices with distance from the platforms.

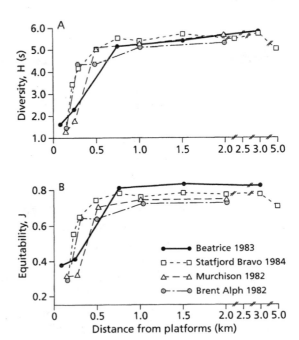

Fig. 7.6 Alterations in values for (A) the Shannon–Wiener diversity index ($H_{(s)}$) and (B) Pielou's equitability index (*J*) with distance from four North Sea oil production platforms. After Kingston (1987).

The validity of relatively simple statistical techniques used to produce diversity indices has been the subject of considerable debate in ecology, and more thorough analyses, involving multivariate statistical methods, have proven to be much more useful. Through use of these methods, patterns of change in communities due to chemical contamination have been accurately depicted.

Field-based studies of the ecological effects have long been recognized as the 'ultimate' evidence that the presence of contaminants in an environment has repercussions at the highest levels of organization. However, relating ecological effects to specific contaminants is very difficult indeed, and remains one of the great conundrums in ecotoxicology. In order to improve our understanding of the effects of toxic contaminants upon communities, another quite different approach has been used in recent years. This involves 'mesocosms'— multispecies systems in which chemical (and physical) manipulation of the environment can be controlled, and observations can be made on the communities present (these are described in Chapter 5). Such studies may be undertaken on a small scale ('microcosms'), on a large scale ('macrocosms') or, more commonly, on a medium scale ('mesocosms'). Both laboratory and field experiments have been undertaken using this approach, with the objective of studying the fate of pollutants in 'ideal' communities and relating this to biological effects. The premise of these studies is that they will mimic, on a miniature scale, the effects of contamination in large communities or ecosystems. Because the mesocosm approach uses many species at the same time, its value is considered to be higher than studies using single species. However, this approach is, by definition, very difficult and expensive to undertake, and its real worth to ecotoxicology remains a subject of continuing debate.

Conclusions: estimating the hazards of chemicals to an environment

A multitude of techniques are available in environmental monitoring and assessment of the hazards of toxic chemicals, but no one technique can provide all the answers. The only alternative is to use a suite of methods to provide the best possible, and most comprehensive, appraisal. There are many examples of multiple techniques used in environmental monitoring—for example, the use of chemical data, scope for growth and condition indices cited above. The State Musselwatch of California (see Box 7.1) has also taken this approach, using a combination of the analysis of contaminants in mussel tissues, scope for growth and laboratory bioassays using a suite of toxicity tests based on local organisms in order to provide a comprehensive examination of marine contamination and its effects. Such an approach has also been extended to the area of marine sediments, which can be particularly troublesome in certain areas close to large cities, where dredging and removal of sediments can cause profound localized effects.

In this area, the 'sediment quality triad approach' has been used. This utilizes three separate pieces of information:

1 the chemical analysis of contaminants in sediments;

2 bioassays of sediment toxicity in a battery of different tests (both lethal and sublethal); and

3 ecological studies involving benthic community structure.

Strong correspondence has been shown between these various measures in the triad which has resulted in locally based but accurate assessments of sediment quality. These assessments can be used to quantify sediment toxicity characteristics and the likelihood of effects if sediments are dredged and removed to another locality. These studies reinforce the notion that any environmental monitoring—be it in terrestrial or aquatic environments—relies on a variety of techniques and assessments in order to provide worthwhile information to environmental management agencies.

The importance of field evaluations of toxicity in environmental assessment is considerable. Certainly, an increasing reliance is placed in modern ecotoxicology upon biomarker measures, which are often performed in laboratory situations, but field-based measures, including ecological surveys, also play an important part. However, for the latter, it is very important to establish baseline or reference sites which can be adequately compared with contaminated areas. In recent times sites which are increasingly affected have been chosen as controls, hence 'shifting the baseline' of comparison, and thus compromising environmental monitoring data.

Risk assessment has become an important technique and this relies on evidence from multiple approaches. Those most commonly used are listed below.

1 Chemical measurements of various environmental compartments, including waters, air, sediments and biota. Choices need to be made amongst these based upon the most likely sources of valuable data (e.g. waters vs. bioaccumulation).

2 Measurements at the level of the individual organism (be it at the subcellular, cellular, tissue or individual level).

3 Ecological data, at the population level where applicable, but more likely at the level of communities. Although this is the most difficult level at which to interpret ecotoxicological data, it is the level where 'ground-truthing' of data obtained from lower organizational levels is required, and the area where, ultimately, the effects of contamination may be felt most severely.

Both biological and chemical data are required to estimate the hazards caused by toxic chemicals in the environment.

Further reading

Depledge, M.H., Amaral-Mendel, J.J., Daniel, B. *et al.* (1993) The conceptual basis of the biomarker approach. In: *Biomarkers. Research and Application in the Assessment of Environmental Health* (eds D.B. Peakall & R.L. Shugart), pp. 15–29. Springer-Verlag, Berlin.

Gilbert, O. (1965) *Symposia of the British Ecological Society* **5**, 35–47.

Hopkin, S.P. (1993) *In situ* biological monitoring of pollution in terrestrial and aquatic ecosystems. In: *Handbook of Ecotoxicology* (ed. P.W. Calow), Vol. 1, pp. 397–427. Blackwell Scientific Publications, Oxford.

National Research Council (1987) Committee on biological markers. *Environmental Health Perspectives* **74**, 3–9.

Peakall, David (1992) *Animal Biomarkers as Pollution Indicators.* Chapman & Hall, London.

Phillips, D.J.H. (1980) *Quantitative Aquatic Biological Indicators. Their Use to Monitor Metal and Organochlorine Pollution.* Applied Science Publishers, London.

Phillips, D.J.H. (1993) Bioaccumulation. In: *Handbook of Ecotoxicology* (ed. P.W. Calow), Vol. 1, pp. 378–396. Blackwell Scientific Publications, Oxford.

Phillips, D.J.H. & Rainbow, P.S. (1989) Strategies of trace metal sequestration in aquatic organisms. *Marine Environmental Research* **28**, 207–210.

Walker, C.H., Hopkin, S.P., Sibly, R.M. & Peakall, D.B. (1996) *Principles of Ecotoxicology.* Taylor & Francis, London.

8: Ecological Risk Assessment

Introduction

The concept of evaluation of risk has its origin in the insurance industry in the late 1600s. At that time data was becoming available on the life expectancy of and the rate of death in the human population. This made possible a rational evaluation of the economics of accepting premiums as payment for the insurance of human lives. Later data on the occurrence of fires, accidents, shipwrecks and so on allowed this approach to be extended to other areas of insurance.

In recent years this quantitative approach has been extended to evaluations of risks to human health. Thus risk analysis of chemical residues in food, water, soil and the air have been carried out. The objective has been to quantify the risk to health so that management policies and procedures can be placed in a quantitative framework and resources can be used in the most effective manner. Human health risk assessment has a basic framework which consists of firstly identifying the hazard involved followed by an evaluation of the exposure to this hazard as well as an assessment of the dose–response relationships for the chemical hazard. Finally an integration of these two factors of exposure and dose response can give a characterization of the risk in particular situations. The risk characterization is then followed by communication of the risk to involved parties and development of risk management strategies. This experience has been drawn on to develop *ecological risk assessment* (ERA) procedures but ERA has distinctive features directly related to ecosystems rather than to human health.

ERA is concerned with the prediction and evaluation of the effects of chemicals, and often other stressors, on ecosystems, usually in specific environmental management situations. It is more complex than human risk assessment because it addresses the effects on whole ecosystems which consist of many species at many levels of organization and complexity. It is based on ecotoxicology which is concerned with understanding the effects of toxic substances on ecosystems. Thus ecotoxicology is a scientific discipline with a conceptual framework related to the organization and development of knowledge. However, ERA differs from ecotoxicology since ERA is directed towards management objectives and is related to human society and the human concept of the hazard. ERA may be needed in a variety of management situations to guide and manage a human-related change leading to exposure of an ecosystem to a toxic agent. In this chapter we are concerned with exposure to chemical agents but in general ERA can be applied in other situations as well, such as development of domestic housing, location of industry and so on. Often ERA is associated with industrial, or other, discharges to aquatic ecosystems, the clean-up and remediation of contaminated sites, development of new pesticides and so on. Government agencies at a variety of levels from local to national may require ERAs to be carried out on activities within their jurisdiction. The basic procedures used in most areas are structurally similar but may differ in the detailed application of these principles. In addition, the ecosystems and environmental conditions differ from area to area leading to differences in applications in detail. Thus ERA may yield different outcomes in these different situations and environments. ERA is in a state of rapid development, particularly under pressure from government agencies to provide accurate assessments and advice which can be used in policy formulation in environmental management. It can be expected that the procedures will be further developed and improved in the future.

Ecological risk assessment

ERA can be performed for a variety of reasons and these may influence the nature of the procedure used although the principles involved are much the same in all situations. Some of the situations requiring ERA with chemicals are: the assessment of discharges from industry and waste disposal areas; the assessment of the effects of chemicals such as pesticides in potential usage situations; and the assessment of effects of spills and other incidents on affected ecosystems. Other stressors may also be evaluated by the ERA procedure apart from chemicals. ERA evaluates the possible occurrence of adverse ecological effects in as quantitative a manner as possible.

ERAs can be considered to be one of two basic types related to the nature of the process involved. In this respect ERAs can be either *predictive* or *retrospective*. Predictive ERA attempts to predict the effects of introduction of the chemical agent in new situations based on the procedure previously described. This is carried out prior to the discharge occurring and the results of the ERA used as a guide to management and in many cases the design of the activity resulting in discharge. Retrospective ERA involves an evaluation of discharges from an existing activity. This may be part of an ongoing evaluation programme to ensure that environmental management procedures for the discharge are effective, or it can be the result of a change, possibly an increase, in the discharge being under consideration. Retrospective ERA can be more accurate and reliable than predictive ERA because it can be based on an evaluation of current effects in the existing situation and, starting from that base, can extrapolate into the future. Clearly, the effects in the existing situation are a more reliable basis for extrapolation than the use of information from models and the scientific literature in laboratory experiments.

While ERA is generally more complex than human risk assessment the technique has some considerable advantages over the human health risk

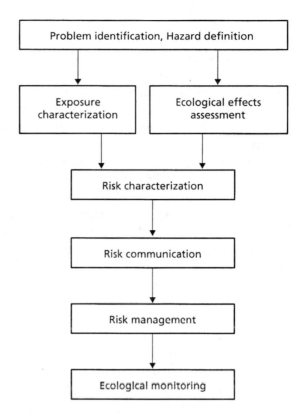

Fig. 8.1 The conceptual framework for ecological risk assessment including risk communication and management and ecological monitoring.

assessment counterpart as well as some significant comparative disadvantages. In human health risk assessments it is extremely rare for the toxicological information available to be actually derived from experiments on human beings themselves since experiments on humans are usually not permitted. On the other hand, in ERA experiments can be conducted on organisms from the natural ecosystems. For example, if a particular population of fish or other organisms is under particular threat then these organisms can, in most cases, be tested in the laboratory and appropriate information derived which can be used in the ERA procedure. However, there remains the considerable disadvantage in that numerous species are involved in ERA at many different stages of the life cycle and that complex interactions between these species occur and can affect the final effect of chemicals in an ecosystem.

ERA has a similar structure to health risk assessment. Health risk assessment consists of a series of sequential steps: hazard identification; exposure assessment; consideration of dose–response relationships; risk characterization; risk communication; risk management; and finally monitoring. In most applications of ERA the steps shown in Fig. 8.1 are used. It can be seen that ERA is a structured approach which allows the systematic evaluation of possible ecological effects resulting from chemical stress. Initially, the problem which is of concern needs to be clearly formulated. The sources of the chemicals and the types of ecosystems exposed are identified. This should allow the ecological endpoints to be specified and the chemical hazards involved to be identified. Exposure assessment is carried out on the contaminants identified as hazards by using models, chemical analysis and so on. Ecological effects assessment attempts to establish the relationship between ecological effects and the degree of exposure to chemical agents. This process usually uses the most prolific source of information on adverse chemical effects which is the LC_{50} data set on various species. This information is then used to extrapolate to the species in the ecosystems. In recent years increasing use has been made of the lowest observable adverse effect level (LOAEL) and the no observable adverse effect level (NOAEL), since these measures are more relevant to the levels of contaminants that are likely to occur in environmental situations. The risk characterization results in an integration of the exposure characterization and the ecological effects assessment to give evaluation of likely effects in the ecosystem under evaluation at the concentrations likely to occur in the contaminant.

The formulation of the ecological endpoints, mentioned previously, will normally involve a range of people and groups from environmental managers to community groups. It should be recognized that this is a sociopolitical process which reflects the human values placed on the system rather than any inherent ecological characteristics. In addition, it may also involve communication with the general public to allow feedback from this important source of information as well. When a risk characterization is complete then communication and interaction with all involved parties is necessary. This may require a review of the procedure identified in Fig. 8.1 potentially starting at the beginning with problem identification and hazard definition, although it could start at any appropriate stage or may not be required at all. As a result, a risk management programme is eventually determined which may be implemented by government, industry or community groups or combinations of these groups. To evaluate the success of the risk management programme an ecological monitoring programme is needed. It would be expected that this would be based on the ecological endpoints identified earlier and extend over a specified time period. After a reasonable time period has elapsed in which the ecological monitoring is able to reveal the state of the ecosystem the results of the monitoring programme can be fed into the ERA process at any appropriate step shown in Fig. 8.1 and the process reactivated and repeated, possibly leading to a modified risk management and ecological monitoring programme.

Problem identification and hazard definition

Clear identification of the problem and definition of the hazards involved is essential for insuring a successful ERA. There are many problems which could be associated with the situation resulting from a particular development or activity. Also the way in which an ERA will assist in the management process needs to be a foremost principle guiding the process. Initial considerations may be concerned with the possible sources of chemicals. From considerations of the type of industry and scale involved, estimations can be made of the waste-water discharges, gaseous emissions and so on. The quantities of the emissions and the possible composition of them can be used to estimate chemicals in resulting discharges which may occur. Some of these chemicals may already be subject to government regulation regarding the level of discharge or the ambient levels which can be tolerated.

Also foremost amongst considerations in ERA is the identification of the specific geographical area involved, including the size of the area and the specific zones which may be important. Zones could be of aquatic systems, terrestrial systems or zones of specific significance, for example, national parks. Consideration of the zones of concern is likely to be influenced by social and political factors as to which values the community places on various zones and ecosystems. Conservation groups, fishermen, hunters and others are all likely to have a different view on the ecosystem features which are of value.

Other factors which need to be considered and taken into account are existing land ownership, industrial operations, land usage patterns and so on. In addition, practical factors are also important. These include such matters as the resources available to actually conduct the assessment, existing knowledge of the area, the expertise available and so on. Another important practical consideration is the actual time available to carry out the ERA before a decision is required. So a wide range of social, economic, political and ecological factors need to be taken into account in identifying the problem and defining the hazard involved.

Ecological effects assessment and ecological endpoints

It is not possible to evaluate the effects of a chemical on all of the components of an ecosystem. It is far too complex and the knowledge base for carrying out such an evaluation is inadequate. Thus in order to practically assess ecological effects an evaluation of selected ecological components is carried out in terms of defined ecological endpoints. Ecological endpoints, evaluated in terms of representative parts of the biological response of the system, from the level of organism through to ecosystem, are shown in Fig. 8.2. An ecological endpoint could be based on a single species, for example, a rare species; a functional group of species, for example, raptors; or an ecosystem function, for example, nutrient cycling. Similarly an ecological endpoint could be based on a specific type of habitat or an area regarded as particularly important. Ecological endpoints are a clear description of specific ecological components at risk and an attribute which defines the level of that risk. For example, a sport fish and a measure of the size of its population or a rare and endangered species and a measure of the numbers of breeding pairs. In some cases whole ecosystems can be defined by using a measure of species diversity.

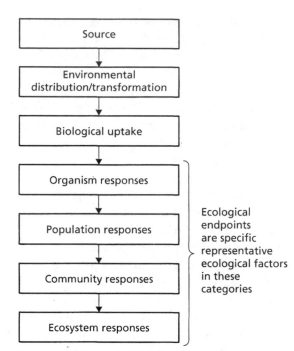

Ecological endpoints are specific representative ecological factors in these categories

Fig. 8.2 Relationship of ecological endpoints to the effects of a chemical on an ecosystem.

The identification of ecological endpoints is a complex task involving such factors as social relevance, ecological relevance and economic relevance as well as the ability to define and measure the characteristic in an unambiguous way. Ecological endpoints defined in this way should express management goals for the ecosystem. If they do not represent important management goals then the value of ecological endpoint is questionable. A danger in this procedure is defining the ecological endpoint in too broad a manner. Many may argue that concern is not with the specific factors which are considered as ecological endpoints, as above, but in the integrity and function of the whole ecosystem. While this is a reasonable concern the breadth of the requirement may not provide a suitably specific basis for a quantitative evaluation. If this occurs such an ecological endpoint is not useful. Alternatively, several well-defined endpoints can be used, which are representative and indicative of the functioning of the whole ecosystem. Of course, such selected endpoints can only be considered to be representative of the whole ecosystem and the success of such an approach depends on the way in which the representative endpoints are chosen. On the other hand, too narrow a definition of ecological endpoints may not allow the function of the ecosystem to be properly evaluated. Such narrowly defined ecological endpoints can fail to include critical variables which are essential for the assessment of ecosystem function.

In precisely defining ecological endpoints which can be used in given situations, spatial factors must be considered. Biological biota in any given area are subject to variable distribution patterns. Different distribution patterns related to environmental factors and the possible spatial effects on species distribution need to be considered in defining the nature of an appropriate endpoint. Also, in the life cycle of an organism there are many stages with different susceptibilities to chemicals. This means that the nature of any indicative organisms will need to be considered in terms of the stage of the life cycle with which we are concerned and possibly other factors.

Exposure assessment

Exposure assessment is concerned with the exposure of the specific environmental receptors to the toxic agent in the ambient environment. The general concept of ERA is diagrammatically represented in Fig. 8.2. Sources are identified and the chemical composition associated with them quantified. In this way the discharge of specific chemicals into the system can be quantified. In addition, the timing of discharges on a daily, seasonal or annual basis may be important. It should be remembered that natural sources can be important with some toxicants, such as metals, and must be taken into account with exposure. Also, in some cases existing environmental contaminants, such as the polyaromatic hydrocarbons, may be important because these substances are widely distributed in the contemporary environment. Thus pre-existing contamination may lead to background contamination and this may need to be considered.

The environment can be characterized as a set of phases with chemical components being distributed between these phases as described in Chapter 3. During the process of distribution, and after the distribution is complete, degradation occurs. Oxidation, due to the presence of 20% oxygen in the atmosphere and high levels in other environmental components, leads to the formation of oxidized and other chemically modified products which generally have a lower toxicity but not always. Reaction with water, hydrolysis, can occur due to the presence of water in oceans, streams, lakes and the atmosphere and, in fact, all sectors of the environment as well as biota. Generally products are formed which are more polar and water soluble so they can be excreted and removed. Finally, in its distributed, degraded and transformed state the toxicant comes into contact with the biological system.

The exposure pathway describes the route by which a chemical agent moves from its source to ecological receptors within the environment. A possible set of exposure pathways for fish from an aerially applied pesticide is shown in Fig. 8.3. The physicochemical properties of the chemical are important in influencing the distribution patterns within the environment. For example, vapour pressure controls the vaporization of the substance into the atmosphere and water solubility influences the movement of the chemical in aquatic areas. Also, the octanol/water partition coefficient (K_{ow}) is often used as a measure of the possible bioaccumulation of a compound by aquatic biota. There are various models available for calculating the distribution of a chemical in the environment. One type of model which is widely used is described in Chapter 3 and is referred to as the *fugacity model*.

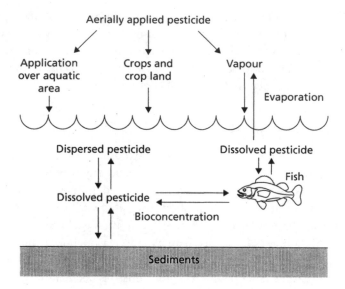

Fig. 8.3 Pathways of distribution of a pesticide in an aquatic area.

The exposure to the toxicant is usually characterized as the environmental concentration in contact with the ecological receptor. Alternatively exposure can be expressed as the amount taken up and deposited internally at the target site within organisms where toxic expression is initiated. This is a less common approach but can be useful and would be expected to more accurately reflect the toxic response since the toxicant is at, or near, the site of action.

The duration of exposure is also an important variable influencing the toxic response. This can be expressed as the period during which a toxic threshold is exceeded. The spatial dimension of exposure needs to be quantified as well since the concentration of a toxicant usually declines with distance from the source. An exposure profile is often developed that can be constructed with dimensions of intensity, duration and spatial extent.

Risk characterization

Risk characterization is an evaluation of possible ecological effects in terms of the identified ecological endpoints at the calculated exposure levels. Thus it represents an integration of the ecological endpoints and the environmental exposure to give an overall evaluation of ecological effects. An ecological response analysis may also be carried out which may be an evaluation of ecological effects with varying exposure levels. Of course this would require that sufficient information on biological responses be available for this evaluation. So in essence the risk characterization is concerned with the use of dose–response data, or derived data, in terms of the ecological endpoints and the calculated exposure levels.

Of course risk evaluation is best done by evaluating the actual ecosystem under investigation. However, this is often not possible and the evaluations usually have to be based on laboratory data on toxicity effects and extrapolation to actual situations. This poses several limitations which must be kept in mind. For example, exposure conditions in the laboratory and the field may be different in terms of such factors as temperature, light, salinity, dissolved oxygen concentrations and so on and thus the conditions under which the toxicity information is derived differ from those existing in the field. This can be overcome by the use of *uncertainty factors*. An uncertainty factor is a factor by which the expected toxicity is lowered to account for unknowns or uncertainties in the original data set. So an uncertainty factor may be used to account for the uncertainty in terms of environmental conditions mentioned above.

In almost any environment there will be insufficient data on the toxicity of chemicals to the biota in the system to allow a complete risk characterization to be conducted. Existing data on toxicity derived from laboratory experiments can be extrapolated in a variety of ways. Firstly, the data on specific chemicals can be extrapolated to other chemicals by the use of techniques such as quantitative structure activity relationships (QSARs) which are described in Chapter 3. In addition the data on biological groups with similar features may allow extrapolation from one taxonomic group to another group for which data is not available. Thus in this way the toxicity data base can be expanded and extended to cover chemicals and biota not currently available.

The information gathered in evaluating the toxicity in terms of the ecological endpoints and the potential environmental concentrations can be expressed as a toxicant response profile which summarizes all of the information available on the ecosystem response at calculated exposure concentrations in terms of the identified ecological endpoints.

Risk communication and risk management

The evaluation of the technical factors in risk assessment is usually carried out by specialists in the various fields involved. However, the fundamental objectives of the evaluation of ecological effects and definition of the ecological endpoints involve a wide range of groups. The initial definition of the appropriate environmental endpoints should involve groups ranging from environmental managers to community groups. Similarly, the outcome of risk characterization needs to be evaluated particularly with environmental managers and other appropriate groups who have input into this process. The communication of the risk identified to groups unfamiliar with the concepts involved can be a complex and challenging task requiring the skills of specialists familiar with ways in which this can be done. It should be recognized that risk communication is a two-way process. An environmental management programme which has aspects that are important to many groups including conservation groups, professional fishermen, environmental managers and so on may often be required. Thus the input from these groups needs to be considered by risk managers as appropriate.

Alternatives for risk management need to be explored with the broad range of groups identified previously. It is clear that in this process there is the potential for a great deal of conflict between the various parties involved such as community conservation groups, industry, professional fishermen and so on. In addition there is a need to consider a host of additional factors at this stage including cost/benefit of alternatives, human health assessment, legal constraints, technical feasibility of alternatives and so on. After consideration of all the technical, social, economic and other factors then a risk management plan can be developed.

Monitoring programmes

When a management plan has been put in place a programme to evaluate the ecosystem response and success of the plan is needed. The ERA process has many areas where errors and uncertainties can occur, ranging from the evaluation of exposure to the identification of the ecological endpoints. Monitoring programmes need to focus on the identified environmental endpoints which are related to management requirements and should reveal the accuracy of predictions and the appropriateness of decisions made concerning management. There is also the need to monitor somewhat more broadly to ensure that unexpected outcomes are detected.

There are two basic types of monitoring programme. Firstly, there are programmes which are concerned with environmental monitoring. These programmes identify and measure the ecological impacts from developments using repetitive observations over a period of time according to a specifically designed sampling programme. These programmes can be, at least partially, focused round the identified ecological endpoints but not exclusively since the endpoints in themselves may not truly reflect processes occurring in the whole ecosystem. Thus, these programmes should be used to test hypotheses about the potential impact of a development on the whole ecosystem. The second set of monitoring processes is concerned with how efficiently a project is meeting the environmental and other objectives set for it in the environmental management plan. So this area of monitoring is focused on the development itself and is often referred to as environmental auditing. Essentially, these programmes are concerned with evaluation of the compliance requirements for a development set out in the environmental management plan. Factors which should be examined in this monitoring and auditing process are the effectiveness of the monitoring of discharges and any other commitments to environmental management. In addition these programmes should determine whether mitigation and other management procedures are effective in minimizing any adverse impacts associated with the development.

Conclusions

The ERA process is an important aspect of environmental management and continued government interest in using this technique ensures continued improvement and growth in the area. Currently, the ERA process is not definitive and precise and, considering the complexities involved, it never will be. All aspects involve inaccuracies and uncertainties to a greater or lesser extent, and this situation will continue. As there are many different types of ecosystems, including marine, estuarine and terrestrial systems, each with its own high level of complexity, ERA can only be applied in natural environments in a generic manner even under the best circumstances. Of course the possibility of retrospective ERA may help in application of the technique in specific situations. However, our understanding of the fundamental functioning of ecosystems, without the interaction of chemical stressors, is usually limited. The data base on chemical interactions with biota cannot cover all possible interactions of all chemicals with all biota at all life stages under all environmental conditions. This means that the available data base will always need to be expanded by extrapolation techniques which will extend our knowledge of both biological responses and chemical interactions.

The complexities of numbers of chemicals and ecosystems have been mentioned previously. But in addition another level of complexity relates to the chemical toxicants which can occur either as single compounds or as mixtures of an infinite array of complexity. In most situations, complex mixtures of chemicals occur in discharges and must be accounted for in the ERA process. In this area there are particular problems in allowing for synergistic and antagonistic interactions and biological responses resulting from particular chemical interactions in mixtures. This is one of the most challenging areas in ERA.

The ERA process provides a framework for the evaluation of chemical effects on ecosystems. It structures the process into a logical sequence and allows for a systematic evaluation to occur which makes the best use of the information available. In addition to providing the information on ecological effects the process identifies areas where new information and development is required. This will allow the expansion and development of the ERA process to occur in an effective manner where most value is obtained from the limited resources available.

Further reading

Cairns, J. (1998) Endpoints and thresholds in ecotoxicology. In: *Ecotoxicology* (eds G. Schüürmann & B. Markert), pp. 751–768. John Wiley & Sons and Spektrum Akademischer-Verlag, New York.

Suiter, G.W. (1995) Introduction to ecological risk assessment for aquatic toxic effects. In: *Fundamentals of Aquatic Toxicology* (ed. G.M. Rand), pp. 803–816.

US Environmental Protection Agency (1996) Proposed guidelines for ecological risk assessment, part II. *Federal Register* **61**, 47551–47631.

9: Ecotoxicology and Management of Chemicals

Introduction

To date, more than 2 million types of commercially produced chemicals exist, and this number is rising by at least 2000 every year. Most commercial chemicals are produced for the benefit of mankind in the form of manufactured goods, food additives, therapeutic drugs and pesticides. However, the presence of these chemicals in the wrong place at the wrong time can result in serious environmental and/or health problems. Thus there is a clear need to understand, assess, monitor and predict the potential hazard of existing and new chemicals to both humans and the environment. Many countries now have legislation requiring the testing and assessment of toxicity and ecotoxicity of all new chemicals, as part of a comprehensive risk assessment procedure, before they can be made available for use.

Ecotoxicology can make a significant contribution to the protection of biological systems by providing a basis for assessment of the potential adverse effects arising from commercially produced chemicals when they enter ecological systems. Ecotoxicity tests aim to provide information on the potential of chemicals to cause harm and the extent to which the adverse impact is likely to be realized in real-world scenarios. In order to ensure objectivity and consistency in decision-making it is necessary to develop and make use of a set of well-defined decision rules.

In Europe, the approach adopted for the control of hazardous substances is to manage risks at a level as low as reasonably practicable. For the control of environmental pollution, the main objective is to use the best or most cost-effective techniques available. This recognizes the need for a balance between costs and benefits of any proposed control measures. In addition, it is important to note that sufficient scientific knowledge may not always be available to provide clear and unequivocal answers to questions concerning the potential costs and likely benefits. On this basis, the community and/or government will have to decide whether it would be appropriate to adopt the precautionary principle. The precautionary principle, previously termed 'anticipatory protection principle' urges greater caution in the anticipatory phase of environmental perturbation. It argues that potentially damaging environmental perturbations should be prevented even when there is no scientific evidence to prove a causal link between causes and effects. It focuses on the need for more effective preventive action and the introduction of control measures without demanding proof of causality between environmental stressors and their effects.

In the US, two major governmental bodies are responsible for regulating hazardous chemicals. The Office of Pollution Prevention and Toxics of the USEPA is responsible for managing risks of pollutants and chemicals to the environment and the general public, while the Occupational Safety and Health Administration of the Department of Labor is concerned with hazard assessment and management in the workplace. Some of the major mechanisms and factors involved in the management of toxic chemicals are outlined below.

The Toxic Substances Control Act (TSCA)

The Toxic Substances Control Act (TSCA) of the United States of America was promulgated in 1976 to regulate the production, processing, use, transportation or disposal of certain industrial chemical substances and to institutionalize the testing and control over the use of these chemical substances in order to protect human health and the environment. There are 31 sections in the TSCA, and sections 2 and 4 are concerned with the development of aquatic toxicity, bioconcentration, or chemical fate data for existing chemicals. Specifically, section 2 acknowledges that humans and the environment may be subjected to unacceptable risks posed by chemicals, and calls for an assessment of the potential hazard of these substances to health and the environment. It also stipulates that the responsibility for the development of the data bank for this assessment lies with the manufacturer. In addition, the US Congress established, under section 4, the Interagency Testing Committee (ITC) to screen, prioritize and recommend existing chemical substances for testing. The Administrator of the Environmental Protection Agency (EPA) can, under this legislation, require chemical manufacturers or processors to test their products and develop an adequate data set for use by government and commercial organizations, both inside and outside the US. These data sets are essential for hazard and risk assessments. Section 4 also requires that tests on a particular chemical be conducted if:

1 there is reason to believe that the chemical may present an 'unreasonable' risk to human health or the environment, or the chemical is produced in large enough quantities that it may result in 'substantial' human exposure or environmental release; and

2 there are insufficient data to reasonably determine or predict the potential impact of its production, processing, distribution, use and/or disposal.

Although these criteria appear to be straightforward, they are not so straightforward when one tries to apply them in practice. In the context of trying to ascertain whether a chemical may pose an 'unreasonable risk', there may be legitimate concerns when:

1 there is insufficient toxicity data;

2 there is empirical evidence that toxicity of the chemical increases with exposure;

3 compounds with a similar structure are known to cause environmental problems and/or adverse health effects;

4 compounds have high environmental persistence;

5 there are data to suggest that the chemical has high potential to bioconcentrate, resulting in significant toxic effects in organisms of higher trophic levels; or

6 there is information that the chemical can cause serious harm to key life-history attributes (e.g. growth, reproduction) which may ultimately result in an adverse impact upon higher levels of ecological organizations.

Another concern lies with chemicals that may be produced in small quantities (e.g. less than 5000 kg per annum) but, because of the way they are applied, are likely to be released into and transported in the environment and come into contact with humans.

In general, chemicals with accurate and reliable EC_{50} or LC_{50} values greater than or equal to 1000 mg L^{-1} or values equal to or greater than 1000 times the concentrations that are likely to occur in the environment are generally considered as being 'low risk', and thus can be accorded a lower priority. Furthermore, it is suggested that the acute toxicity data should be collected from tests conducted with at least three phylogenetically unrelated species, commonly an alga, an invertebrate and a fish. These directives underline the important contributions ecotoxicology can make in the protection of the environment and its human inhabitants. In this regard, good quality ecotoxicological data is not only necessary, but essential. Consequently, most testing protocols now stipulate, apart from general good laboratory practice (GLP), other testing conditions and requirements, such as the need to measure chemical concentrations before, during and after test duration, and whether static or flow-through exposure systems should be used. It is clear that standard testing protocols are needed which are validated and accepted by regulatory and testing agencies in different countries, such as the EPA, the American Society for Testing and Materials (ASTM) and the Organization for Economic Cooperation and Development (OECD). This is necessary to achieve the overall objective of protection of ecosystems without entailing excessive cost.

Control of chemicals in Europe

Since 1981, more than 5 million chemicals have been listed in the Chemical Abstracts produced by the American Chemical Society as a comprehensive record of chemical research. It is probably reasonable to assume that chemicals not placed on the market are unlikely to have a significant impact on man and the environment. Nevertheless, over 600 new chemicals have been notified under the new chemicals notification scheme of the European Community in that same period. Under this scheme, information relating to the identity and toxicological and physicochemical characteristics of these new chemicals is gathered before they can be made available to the public. It is useful to note

Table 9.1 Requirements of the European Chemical Notification Scheme.

Level	Volume of production	Requirements
0	1–100 tonnes per year	Identity of manufacturer and notifier Identity of the substance Chemical Abstract Service (CAS) number Chemical composition Impurities and analytical methods Scope of use and anticipated production levels Precautions for use, emergency measures and packaging Physicochemical properties (e.g. vapour pressure, surface tension, water solubility, K_{ow}) Toxicological data (acute toxicity, mutagenicity and carcinogenity) Ecotoxicological data (e.g. 96-h LC_{50} for fish; 48-h EC_{50} for *Daphnia* immobilization test; 72-h IC_{50} for algal growth inhibition assay) Inhibition tests for bacterial activity (measure of biodegradability)
1	100–1000 tonnes per year per notifier or 500 tonnes in total	Additional physicochemical studies (e.g. development of analytical detection methods for transformation products) Additional toxicological data from fertility studies, additional mutagenesis studies and toxicity studies Additional ecotoxicological data from algal tests, and studies investigating chronic toxicity on fish and *Daphnia*, and species accumulation
2	Over 1000 tonnes per year per notifier or over 5000 tonnes in total	Additional toxicological data from chronic toxicity tests, additional studies on carcinogenity, development toxicity and teratogenicity Additional ecotoxicological data from tests for accumulation, degradation, mobility and absorption/desorption; toxicity studies on birds and other organisms

that the notification procedure only deals with pure substances, and does not handle preparations composed of several chemicals. However, in the case where a preparation containing a new substance is imported into the European Community, that new substance will have to be notified. The scheme has three levels as shown in Table 9.1.

The notification scheme outlined above is not applicable to the more than 100 000 chemicals that existed in the European Community before the critical date when the notification scheme came into force. These chemicals are listed in the European Inventory of Existing Commercial Chemical Substances (EINECS). This inventory provides for each chemical minimal information such as:

1 an EINECS number;
2 the name of the chemical;

3 the molecular formula; and
4 the Chemical Abstracts Service (CAS) number.

Role of the OECD in the control of chemicals and international standardization of testing methods

The earlier work of the Organization for Economic Cooperation and Development (OECD) on chemical safety was focused on specific compounds including persistent organochlorine pesticides, polychlorinated biphenyls and heavy metals such as mercury, cadmium and lead. More recently, the organization has broadened its scope in terms of chemicals considered and to the general area of protection health of human and the environment. Specifically, the OECD Chemicals Programme has been engaged in setting test guidelines, promulgating good laboratory practice, initiating hazard assessment exercises, and harmonizing classification and labelling systems. The overall objective is to allow free circulation of goods while at the same time ensuring an adequate level of protection for human health and the environment.

Some of the driving forces behind OECD's involvement in standardizing testing protocols for new and existing chemicals are as follows.
1 International harmonization of test methods will increase the validity and acceptance of test results by the international community, and thus help to prevent duplication of effort, and reduce the unnecessary use of laboratory resources.
2 Pressure from legislation, customers and other stakeholders requires that for industry to be competitive in the world market-place it is necessary for it to understand its exposure to environmental legislation and its impact on the environment, and to be able to demonstrate credibly and transparently this understanding and its attempts to minimize its adverse environmental effects. Standardization of test methods and protocols for potentially hazardous chemicals will assist in removing barriers caused by national boundaries, thereby facilitating international trade.

To formalize the international acceptance of data on chemicals, the OECD *Decision Concerning the Mutual Acceptance of Data in the Assessment of Chemicals* was adopted in 1981. The gist of the decision was that data generated by OECD member states in accordance with OECD test guidelines and OECD principles of good laboratory practice (GLP) should be accepted in other member states for purposes of assessment and other uses relating to the protection of human health and the environment. This move provided an important bridge between the US (i.e. in terms of the requirements of the TSCA) and the OECD countries (i.e. in terms of the OECD test guidelines) and has allowed the sharing of data between the US and European countries. The following are examples of the OECD ecotoxicological tests used to identify short-term acute effects.
1 Growth inhibition test of algae.
2 Acute toxicity test of zooplankton.
3 Acute toxicity test of fish.

Similarly, specific TSCA requirements for aquatic toxicity and bioconcentration testing have included the following.

1 Acute algal toxicity tests involving species such as green algae (*Selenastrum capricornutum*) and diatoms (*Skeletonema costatum*).

2 Acute and chronic toxicity tests on daphnids (*Daphnia magna*) and gammarids (*Gammarus fasciatus*).

3 Acute toxicity test on fish, e.g. rainbow trout, flathead minnow, bluegill.

4 Early life-stage toxicity test on fish, e.g. rainbow trout, flathead minnow, sheephead minnow.

5 Bioconcentration tests on fish and oysters.

6 Acute and chronic toxicity tests on mysid shrimps.

Assuming that the number of existing chemicals evaluated by the US is between 50 and 100 a year, it is clear that not all existing chemicals can be tested under the present scheme and testing efforts. Consequently, only certain existing chemicals, particularly those suspected to cause potential concern to humans and the environment, are subject to environmental risk assessment—a process by which the risks arising from the production, use and disposal of new and existing chemicals are identified and quantified. In general, full assessment is required if:

1 there are indications of bioaccumulation potential;

2 ecotoxicity tests indicate that the substance is potentially mutagenic, or may cause irreversible harm to the environment or serious damage to health following prolonged exposure; or

3 Data on structurally analogous substances suggest a potential hazard.

In March 1993, the European Community adopted a Council Regulation on the evaluation and control of the risks of existing substances. Under this Regulation, each member state is required to act as a rapporteur for a number of priority substances which have been recommended for risk assessment. Each member will evaluate the information submitted on these substances and is required to prepare a risk evaluation of the substances with regard to human health and the environment. The assessors are expected to reach one of the following three conclusions.

1 Existing information appears to be adequate, and there is no need, at this stage, for further information gathering or controls beyond those already in place.

2 Existing information is inadequate for a rigorous evaluation, and more information needs to be collected, which may require some form of testing.

3 There is reason to suggest that further control measures on the chemicals need to be implemented.

In making decisions on the above, understanding and assessment of the risks and benefits of the chemicals is essential. Decision makers need to balance a number of conflicting demands, as outlined in Table 9.2.

The two scenarios in Table 9.2 represent the two ends of a spectrum. Very often, the optimum level rests somewhere in between. The primary objective of the above approach, analysing the risks and benefits, attempts to locate that

Table 9.2 Actions to be taken in terms of costs and benefits.

Action to be taken	Costs	Benefits
Complete ban of a chemical	Cost of developing alternatives Increased cost due to unavailability of the chemical	Improved protection to environmental and human health
No control	Potential risks to environmental and human health	Continued enjoyment of benefits from the chemical

optimum point by considering and balancing society's judgements about the relative merits of various options.

A complete risk assessment usually comprises the steps outlined in Chapter 8. In general terms ecological risk assessment can be classified into:

1 predictive risk assessment which estimates risks from proposed future actions such as production and marketing of new chemicals, or commissioning of a new discharge/emission source; and

2 retrospective risk assessment which estimates the risks posed by an event that has already occurred.

The primary objective of environmental risk assessment is to predict the concentration of the substance below which adverse effects in the environmental compartment of concern are not expected to occur, i.e. the predicted no effect concentration (PNEC).

In cases where the estimation of a PNEC is not possible, a judgement may need to be made based on a qualitative estimation of the dose–response relationship. The PNEC may be calculated by applying a safety factor to toxicity data obtained from acute and chronic tests, e.g. EC_{50} or NOEC, etc. The safety factor is included to reflect and eliminate the uncertainties associated with the extrapolation from critical toxic concentrations. In principle, the degree of uncertainty (safety factor) tends to decrease with increases in the amount of available toxicity data.

An important piece of information in the ecological risk assessment is the concentration of the substance that is likely to occur in the environment, referred to as the predicted environmental concentration (PEC). Very often, the PEC is estimated by taking into account a number of factors:

1 the level of production, frequency and duration of exposure of the substance;

2 the form in which the substance is produced, imported and used;

3 physicochemical properties of the substance;

4 use pattern and degree of containment; and

5 environmental fate of the substance and its breakdown and/or transformation products.

Characterization of the risk of a chemical for any given environmental compartment would involve a comparison of the PEC with the PNEC, expressed as a PEC/PNEC ratio. This so-called quotient method uses the PEC/PNEC ratio as an index of risk. The principle here is that if the ratio is ≤ 1, the substance is usually considered as having a low risk level, whereas a ratio > 1 would indicate a potential cause for concern.

A risk–benefit analysis has recently been completed on the chemical tributyltin, a commonly used wood preservative and an important component of many marine antifouling paints, by scientists and social scientists in the UK. A brief summary of their findings is given in Table 9.3.

Table 9.3 Summary of factors involved in the management of tributyltin (TBT).

	Arguments for restricting the use of TBT	Arguments against restricting the use of TBT
TBT in antifouling paint	Increased dry-dock activity resulting in the need for more frequent repainting will imply a rise in employment opportunities in the relevant sector Major reduction in the level of TBT through further restrictions on the use in antifouling paints is required to bring significant improvement to the shellfish industry	Significant increase in cost to the shipping industry (estimated to be at least £50 million over 5 years) May drive potential customers to countries with fewer restrictions Unlikely to provide significant benefits to the terrestrial environment
TBT in wood preservatives	Alternative wood preservatives are available for internal application, and therefore tighter control will incur only minimal extra cost Likely to provide significant benefits to the terrestrial environment	A complete ban or severe restriction on the use of TBT may result in manufacturers not using pretreatment technology or using other technically inferior compounds; increased repair and maintenance costs Minimal environmental gains, e.g. to the shellfish industry, as the major source of TBT is antifouling paints
General use of TBT	Voluntary reduction in use may be insignificant Impact on major producers and users likely to be minimal Reduced potential harm to workers Although no clear indication of potential effects on general public, precautionary principle should apply	Voluntary reduction in use expected due to a need to improve the image of the industry concerned Potential harmful effects on workers can be minimized through good operating practice No significant effects identified. Risks unknown but probably minimal

Conclusions

It is worth noting that most developments in ecotoxicology in both Europe and the US have been concentrated in the aquatic environment. Nevertheless, this is probably a reasonable first step given that aquatic systems, particularly streams and rivers, are commonly used for disposing and transporting wastes, while coastal areas are especially vulnerable to environmental perturbation, including chemical stress. Nevertheless, the effects of chemicals on the atmosphere and the soil are also of major concern, and at least equal emphasis needs to be placed on developing the concepts and methodologies in ecotoxicological studies in these systems.

Further reading

Commission of the European Communities (1992) Chemicals control in the European Community. *Directorate-General XI Environment, Nuclear Safety and Civil Protection.* Royal Society of Chemistry, Cambridge.

Rand, G.M. (1995) *Fundamentals of Aquatic Toxicology: Effects, Environmental Fate, and Risk Assessment,* 2nd edn. Taylor & Francis, Washington D.C.

Index